An Inconvenient Future:
Tomorrow's Future Today

by

Robert Byrum

Dorrance Publishing Co
585 Alpha Drive
Pittsburgh, PA 15238
Visit our website at *www.dorrancebookstore.com*

ISBN: 979-8-88925-388-4
eISBN: 979-8-88925-888-9

Table of Contents

Dedicated to: The climate scientists who have worked so hard to inform the World.

Preface

This is not a happy-smiley book and if you are looking for a happy read; it is not. Shocking it may be to some, unbelievable; hopefully to just a few, and disheartening to those who take the time to understand and explore its message.

We have been living for a very long time with challenges to our way of life and our very existence; wars, the threat of nuclear obliteration, disease, and hazards of social upheaval. Until recently these were the major tests we had as a people, a country, and a world, and although they still exist today, we have managed them and prevailed.

Today, humanity faces a new trial, one that most of us knew little about until recently and many are still unaware of the magnitude and significance of what is happening. The changing world is engulfing us in a slow continuing and increasing steam bath, gradually turning the heat up as we swelter and sweat without a cool down switch. It is understandable that we are confused, distrustful and in denial of the reality that is occurring. We are not prepared nor have we been conditioned to deal with these changes. It is difficult to accept because events have increased so slowly in the past. But now they have finally speeded up and we are beginning, in a serious way, to see and experience firsthand their permanent destructive effects advance into our lives.

How do you cope with an ever-increasing warming earth? For those that follow us down the path towards the future that will be the question for the ages.

Other Books by the Author

A Life Well Lived
Words Matter
Extinction

Introduction

It is not in man's nature to take immediate action when confronted with a difficult threating challenge. We tend to wait, to procrastinate to find excuses or denials for delays or non-action. Even in the face of overwhelming sound evidence or immediate need, we hesitate. We tend to think that the future will be as it has been in the past. It's just too hard to change, its inconvenient, too expensive or embarrassing to omit that our long-standing beliefs may be wrong. We vacillate as individuals and as a nation. We may take small steps to pacify ourselves and others to show that we are doing something, but the big steps that are immediately necessary are just too difficult and inconvenient. We think that maybe that is enough that the problem might just disappear if we remove it from our thinking, or set it aside for other mundane interest.

But there comes a time for action when delay makes no sense. Are we there or have we delayed to long? There are those today, who will continue to obstruct, to deny what is obviously happening with the World's climate crisis, with reasoning based on nothing more than partisan politics. It is often said that doing something over and over and expecting different results is the definition of insanity. It is no exaggeration to say that what we and the rest of the World have been doing regarding the potential climate catastrophe fits that definition.

As humans our time frames are too short. We live in terms of today, tomorrow, next week or possibly even next year. We cannot fathom thinking or planning beyond these times because they do not apply to us. It is difficult for us to project how our actions and non-actions today may affect the future in fifty, a hundred or two hundred years or more. We are not wired for that, it's contrary to our nature to be concerned about something we will not personally experience. We will not be here, so why be concerned about something so far in the future?

But our Mother Earth will still be around, she is not going anywhere. Our short-term human time spans mean nothing to her. She has been in existence for 4.5 billion years. We are the new kids on the block and of no consequence. She is forever. We can drastically change her as we have been doing for a long time but we cannot destroy her. We hear people speak of "destroying the earth." It's true, the damage we have been inflicting on her may make it impossible for her to support life as we have known it. But she will endure, we are not destroying her, we are only eliminating ourselves. We heat and rise her oceans, melt her glaciers, destroy her ecosystems but she is resilient; she is used to change. Since her birth she has been in a constant evolution of slow and on- going change. She has experienced and has recovered from many tragedies and extinctions over her long existence. She has suffered multiple mass destructions during which time more than 90% of her life forms ceased to exist. She has a great advantage over us. Time. There will always be time for her to adapt and mend from changing conditions as she has done in the past. Will we have time? The many hurts we are causing her now will heal. Will we?

The present conditions and necessary decisions facing us as individuals, as a nation and the World are such an example. The human world has had a very short history of what we are just beginning to experience but what the earth has suffered many times in its long past. We have never had the need, the time nor the opportunity to evolve and recover or even prevent the present and future destructive excesses we have brought upon ourselves. We do not know how to cope with the increasing extreme heat, floods and drought, the rise and heating of our oceans, massive and recuring fires, severe storms, melting of the world's glaciers and ice caps; they are beyond our comprehension These events that are occurring today are changing Earths environment and will eventually completely revise ours.

For some time, there has been a growing conflict between two environments. Man has developed his toys: his cars, airplanes, TV, the Internet and thousands of other necessary items. It is only in his recent past that he has come to depend on these necessities. A short two hundred years ago they didn't exist but humans managed to survive, reproduce and prosper. At that time

Earths ecosystems were in a reasonable balance; it was a healthy place. Today that balance has changed dramatically, it is no longer healthy. Man has given Earth a lasting and increasing fever.

Earth's environment is entirely different. It has no necessary dependence. It can exist with the degrading of the quality of its air, its water and increased temperatures. Can man? It will not be the same Earth but it will still exist. Both man and Earth are linked together; it can survive without man but man cannot live without a healthy earth environment. This is not to say that man will disappear, he will still be around. Those who are fortunate to survive may have the opportunity, given time, to evolve and adjust to the changing conditions. The big unknown is what will those conditions be like.

We must understand that we are not a unique species. Man has no more significant than other forms of life; he has come to the present not at the climax of creation but as a physical reaction to nature and the environment. Humans are an accident in the natural development of life, not a product of orderly development. We have survived so far, for only 1% of the time that dinosaurs walked the earth. It is easy for man to believe he is the ultimate in creation. That he is a superior creature and that his journey down through the ages of evolution has been completed. It has not and never will be. Evolution is a slow but continuing process that never ends. Consider how we and the world around us have changed in the last fifty, hundred or two hundred years? What habits and necessities have we acquired; what are the changes that have occurred over those very short time periods? How will they change in the next fifty, one hundred or two hundred years?

"Men will come and go leaving behind only tiny scars to mark their passing. This as it has and always will be. They will one day disappear as the mountains have been eroded, the glaciers melted, and the planets rearranged. We are only a speck in time, only an instant in the constant evolution of change." Author Unknown

"We are like butterflies who flutter for a day and think it is forever." Carl Sagan

We have moved from a rural society of self-reliance to an urban world of dangerous populations squeezed tightly together with millions of people sharing limited critical resources in large metropolitan areas. There is over eight billion of us now roving our planet, soon to be ten billion. Our old transport friend the horse has been replaced by hundreds of million vehicles of every type, speeding each day over paved roadbeds that were once productive lands. Our atmosphere has been degraded and today is filled with thousands of moving objects that we have come to depend on. Many of our essential fresh water supplies have been polluted and in numerous locations is unfit for consumption. We are heating our world to excess by the rapid depletion of its inadequate national resources, burning them up as if they will last forever, they will not; they are not infinite.

How many of these changes have been positive for mankind? It is certainly true that countless numbers of us, not all, live in the riche's societies man have ever known. We have an unlimited number of luxuries no one would have possibly imagine a hundred years ago and we are not about to give them up, nor should we. Our perception as to what is necessary in our lives has so evolved that things that were once considered luxuries or of no account in the past have now become necessities that we will not or cannot part with. Because of our increased numbers we have been forced into conformity with the habits of the masses; we like and contribute to these changes.

But have we reached the point of maximum growth in this country and other parts of the world? And if not, how long can we continue with unending development within the limits we face as a people among our diminishing resources? We must recognize that "more is not better." What are some of those limits? The persistent population increases, fresh water supplies, pandemics, diminished resources like ocean fisheries and arable farmlands and certainly other disruptions triggered by higher temperatures due too global warming.

The billions of people wandering around this big rock we call home have increasing diminish its healthy environment. We occupy over 80% of the earth's land and have become its most populous mammals. The U.S. alone consumes 20% of the world's resources while holding only a small percentage of its population.

Can this persist for the foreseeable future without destroying many of its essential ecosystems and necessary reserves? Both can sustain only so much damage and overuse before they become useless. Over ninety nine percent of all the species that have ever live are no longer with us, they are gone forever. The increase in global warming is telling us we have already exceeded our limits to future growth. It will be impossible, as desirable as it would be, for "clean energy" to satisfy all of our current and future energy needs and accommodate future rapid growth while protecting the worlds ecosystems. We must slow down our expenditure of energy and natural resources. Continuing over consumption of fossil fuels will only decrease the limited amount of time we have available to delay an advancing tragedy.

"We stand now where two roads diverge. But unlike the roads in Robert Frost's poem, they are not equally fair. The road that we have long been traveling is deceptively easy, a smooth super-highway on which we progress with great speed, but at its end lies disaster. The other fork in the road-the one "less traveled by" offers our last, our only chance to reach a destination that assures the preservation of the earth." Rachel Carson in her book Silent Spring

Climate Change and Global Warming

Climate change can make you a prisoner in your own house. It can destroy your home and kill you and your love ones. It can destroy your cultural heritage and long acquired memories. Gone are the days of wondering if the climate crisis will affect you, your neighborhood, your ability to earn a living or your future. It is no longer possible to think you are safe because you don't live near the ocean. It is no longer a question of whether climate change is real, or where it is occurring; it's there wherever you are. Until the climate is stabilized there will not be a new "normal climate."

It will reach deep inland, up rivers and upon the ground where you live and work. It will change what food can be grown in areas that are experiencing new weather patterns and how much you have to eat. It is the flooding and burning of your homes and the heat that continues and will not leave on sweltering summer nights. This is the future that we thought would never come, but the future has arrived visibly and invisibility as sure as the sun rises each day. The tragedies that are occurring around the world today are like post cards from the future that has taken them a while to reach more of us where we live.

The world's warming is the greatest issue of our times, actually of all times. It is also the largest issue that we are ignoring. Climate change is already causing dangerous and widespread disruptions to billions of people and countless species. This is not occurring in some remote future; it is today, everywhere. For the most part meaningful public policy to address and adapt to the increasing heating is sorely lacking in our government and governments around the world. Public discussion and action on the climate are quickly moved to the back burner; displaced by politics, inflation, gas prices, voting rights, abortion and other subjects of seemly importance that fade in the magnitude of the destructive consequences to our rapidly changing and warming world.

Numerous people don't understand the difference between short-term "weather" that is happening daily and the "climate" that is measured over a much longer time period of thirty years or more. As I write this, I glance out the window at the snow and the thermometer that's hovering at 30 degrees (below zero) and I can almost hear the cry, "If global warming is real why is it so cold?" Does the cold weather disprove climate change? No, climate change is the global warming that is occurring over a long time. Weather is today. Winter just doesn't go away because average temperatures are rising. It has taken many years of studying daily weather data to understand the global climate and how it is changing. Climate is about the accumulation of long-term trends and it does not occur evenly around the globe. As you add energy to the system, in the form of heat created by carbon in the atmosphere, extremes can occur. Texas can have ice storms at the same time 33 million people are displaced in Pakistan due to heat and flooding; it is never all one thing or another. "Global warming will not stop the seasons, but it is causing long-term trends in winter conditions that are strong and accelerating." Said Jason Smerdon, a climate physicist at Columbia University. "Today, for every two daily record high temperature maximums that occur there is only one daily record low. This is climate change happening. If the climate wasn't warming there would be an equal chance of a daily record high temperature set compared to a daily record low. Over the rest of this century and beyond the number of record highs will increase but there will still be a few extreme cold events," he said.

Since 1896 average winter temperatures across the lower forty-eight states has increased by 3 degrees. Nationally 57 of the U.S. Weather Service Stations have shown a decline in snowfall since 1930. Winter weather is now on average fourteen days shorter, and summers are seven days longer

The New England states are experiencing temperature warming and sea level rise at a rate faster than the rest of the nation. Up and down the East Coast, difficult decisions are going to have to be made in the next fifty or a hundred years. What do we try to save and at what cost, do we rebuild or should we; can we get insurance, do we leave now, where can we go? The science is clear, many oceans front communities cannot and probably should not be saved from the sea level rise and potential storms caused by the climate crisis affecting the planet. Sure, the ocean rises seams slow but unfortunately it will continue to be consistent over the coming centuries. It has already been programed to increase due to today's amount of carbon already in the atmosphere; It can be slowed down but not stopped. About 90% of North Carolina's 300 miles of coastline are bordered by barrier islands that move constantly. They move, growing and shrinking, even disappearing depending on conditions such as sea level rise, erosion, storms and king tides. They are living on borrowed time and will continue to be un-stable.

It is understandable that residents living in these areas are reluctant to re-locate even knowing the hazards they face. Many oceanfront communities spend huge amounts of money to hold back the sea, trying to defend what in time will be ultimately indefensible. Managing the risk of sea level rise will require communities to reconsider their place alongside the ocean. Sea levels are predicted to rise 11 to 13 inches along the U.S. southeastern shoreline by 2050 and considerably more by the end of the century and it will not stop rising then, according to the National Oceanic and Atmospheric Administration. These low-lying shorelines are constantly at risk of hurricanes and storm surges that today can be as high as 12 to 18 feet, causing complete destruction far inland from the coast. The tragic damage and loss of life caused by the most recent storm **Ian**, on the West Coast of Florida, paints a vivid picture of what can occur today; it will be even more destructive in the future as sea levels continue to rise and the frequency and intensity of the storms increase.

The two elements that are the major cause of our current toasting; carbon dioxide (CO2) and methane gas. For over 300,000 years the level of carbon dioxide in the atmosphere had been stable averaging between 230 parts per million (ppm) and 250 ppm as determined from fossil records and ice cores taken from Antarctic glaciers. Today that level has risen to 419ppm and continues to climb. In the past 800,000 years the level of CO2 has never been as high as what the world is experiencing today.

In 2015, a research team drilling in the ice at Allan Hills, Antarctica retrieved ice that was estimated to be 2.7 million years old. Recent drilling in 2019 went deeper to discover older ice that was three million- to five million years old. From these studies, scientists plan to make comparisons between what the Earth's future may look like by comparing today's atmospheric carbon content to that which was trapped in the ice during those ancient times when there was a similar atmosphere. This will advance the knowledge of what to expect in the future for the world and mankind.

When mankind started the development of today's industrial society, two hundred years ago, the level of CO2 in the atmosphere began to rise, starting the warming process that we are experiencing today. Just two hundred years! What of the next two hundred years? In hardly a wink in time man has changed the world. The level of CO2 will continue to increase every year as we continue to ignore the hazards of burning fossil fuels. Can we slow the increase down? It's not going to be just heat waves, droughts, wildfires, and hurricanes the world will face. It's going to affect fresh water resources, food supplies, eco-systems damaged and destroyed, political, social, and economic upheavals, increased migration and wars. The rate of emissions released and the resulting heating can be and must be slowed down but it can't be stopped.

A history of climate change from an article written by Sylvia Dee, Associated Professor of Earth Environmental and Planetary Sciences, Rice University

In the year 1856, long before the current political divide over climate change, an American woman scientist named Eunice Foote documented the underlying cause of today's climate crisis. In a brief scientific paper, she was the first to describe the extraordinary power of carbon dioxide (CO2) to absorb heat, the driving force of today's global warming.

Carbon dioxide is an odorless, tasteless, transparent gas that forms when we burn fires including coal, oil, gasoline and wood. As the sun heats the earth not all of the heat is reflected back into space. Our atmosphere becomes hotter because gases such as CO2, methane and water vapor absorb and capture some of the outgoing heat radiating back into space from the earth's surface. As additional carbons are released and burned, more heat is captured and stored increasing the warming effect.

Foote conducted a simple experiment. She put a thermometer in each of two glass cylinders, pumped carbon dioxide into one and air into the other then set the cylinders in the sun. The cylinder containing the carbon dioxide got much hotter than the one containing air and she realized that the carbon dioxide would strongly absorb heat in the atmosphere. This led her to the conclusion that "if the air had mixed with a higher proportion of carbon dioxide, then at the present, an increase in temperature would result."

A few years later, in 1861, the well-known Irish scientist John Tyndall also measured the heat absorption of carbon dioxide and was surprised that something "so transparent to light" could so strongly absorb heat. He recognized the possible effects on climate, saying "every variation of water vapor or carbon dioxide must produce a change of climate." He also considered that other hydrocarbon gases, such as methane could add to climate change, again saying "an almost inappreciable addition of gases like methane would have great effects on climate."

By the 1800s, just 200 years ago years ago, human activities were already increasing the carbon dioxide in the atmosphere, by burning increasingly amount of fossil fuels principally coal and eventually oil and gas.

The first estimate of carbon dioxide induced climate change was made in 1896 by Svante Arrhenius, a Swedish scientist and Nobel Laureate. In 1896 he calculated that "the temperature in the Arctic regions would rise 8 or 9 degrees Celsius if carbon dioxide in the atmosphere increased to "2.5 or 3 times" from its present level at that time. It appears that his estimate was likely conservative. Since 1900 atmospheric carbon dioxide has risen from about 300 parts per million to today's level of 419ppm as a result of human activities. The Arctic has already warmed by about 3.8 degrees Celsius (6.8F)

Nils Ekholm, a Swedish meteorologist agreed, writing in 1901 that: "The present burning of pit coal is so great that if it continues, it must undoubtedly cause a very obvious rise in the mean temperature of the earth." All of these facts about climate change were well understood over a century ago! Winning slowly while fighting climate change is the same as losing. Later has already become too late.

In 1937, English engineer Gay Calendar documented how rising temperatures correlated with rising carbon dioxide levels. "By fuel combustion man has added about 150,000 million tons of carbon dioxide to the air during the past half century and world temperatures have increased."

In 1965 scientist warned then President Lyndon Johnson about the growing climate risk, saying: "Man is unwittingly conducting a vast geophysical experiment. Within a few generations he has been burning the fossil fuels that were slowly accumulating in the earth over the past 500 million years." The scientist issued clear warning of future high temperatures, melting ice caps, rising sea levels and acidification of the ocean's waters. Did anybody listen, are they listening now?

In has been well over a century since these first early warnings were sounded; more of the ice caps have melted, sea levels have risen and the ocean has become more acidic due to its increasing absorption of carbon dioxide that forms carbonic acid. This has been a critical problem for ocean dwelling organisms. It is continuing and will only get worse as more greenhouse gases are consumed by the atmosphere.

There should be no question today about the conclusion that human-generated emissions from the burning of fossil fuels are causing increasingly dangerous warming of the climate and other harmful effects? Wake up everyone, the evidence is here staring us in the face with our homes being flooded and burned, heat waves roasting us and our forest and agriculture fields, and the increasing lack of the basic necessities for life; fresh water and food, by millions of people around the globe.

It is a fact that reality is now fast overtaking the scientific predictions. The megadrought and severe heat waves in the western United State and other locations around the world. The massive fires occurring in the U.S., Siberia, Australia, Spain, Portugal and France. The historic low water levels in Europe's and Asia's major rivers and some of the major rivers here in the U.S. including the Colorado and Mississippi Rivers. The relentless rains and flooding and the increase numbers and more powerful hurricanes are all forerunners of increasing climate disruptions today and far into the future.

The world has known about the warming risk posed by excessive levels of carbon dioxide for decades, even before cars and coal fired power plants. It took a rare, in her day, female scientist Eunice Foote 165 years ago to warn the world of the basic science of atmospheric heating. Over the past 200 years the level of carbon dioxide in the atmosphere increase by 25%, but just over the last thirty years we are close to a 50% increase. Why haven't we listened more closely?

The National Oceanic and Atmosphere Administration, NASA, The National Science Foundation, the National Research Council and the Environmental Protection Agency all agree "Earth is warming mainly due to the increase in human -produced heat-trapping gases." The claim that human caused climate change is a hoax is false according to these organizations. A long-term lie has been that climate change isn't real, but increasing weather patterns have made that argument harder to justify. Now the deniers are changing to another lying tactic that says that there are no good alternatives to fossil fuels or the alternative solutions are too problematic or too expensive. "In other words, we are stuck with fossil fuels and there are no good alternatives, so burn baby burn," said Jason Smerdon, a professor of climate physics at Columbia University. Such arguments have been for the benefit of the fossil fuel producers who favor profits over working towards a solution. "Climate disinformation has always been about delaying any action on global-warming," he declared. In fact, this new misguided arrogance is again false and dangerous. Julio Friedman, chief scientist at Carbon Direct, a carbon management firm and former climate professor at Columbia University, states, "we have the technology we need to abate this crisis; it is no longer a question is this really possible, but instead, how quickly can we do it."

The doom proclaimers still insist that it is impossible to reduce greenhouse gas emissions without devastating the economy and reducing our standard of living. But in reality, the technology to decarbonize much of the electrical grid already exist. Today wind and solar development along with battery storage are increasingly cheaper than coal and natural gas. An Oxford University study released in September 2022 found a fast transition to decarbonize energy systems is cheaper than a slow one or no transition at all. It states that achieving zero carbon energy systems is "possible and profitable" and will save the world at least $12 trillion compared with continuing the current levels of fossil fuel use.

The climate disaster has arrived with the souring temperatures across Europe and the U.S. In much of the northern hemisphere the scorching summer temperatures are becoming the normal pattern. As scientific predictions are becoming the reality, the emergency is indisputable and widespread and dramatic weather events are recorded with an ever-increasing frequency. These patterns will have disastrous and far-reaching effects for the natural world, global food supplies, health infrastructure and more going forward. The United Nations President, Antonio Guterres has likened the crisis as "collective suicide."

The world-wide slow down due to the Coronavirus Pandemic in 2020, reduced the yearly release of carbon emissions by about 7%. This was a record decrease but it wasn't even a blip compared to the total carbon dioxide released into the atmosphere that year and had no measurable effect on reducing ocean heating. "The fact that the oceans reached yet another new level of warming in 2020 despite a record drop of 7% in carbon emissions points out the reality that the planet will continue to warm as long as we persist to emit carbon into the atmosphere." Said Professor Michal Mann, at Pennsylvania State University. "It is a reminder of the urgency of bringing carbon emissions down dramatically over the next few years." Ocean heat studies are performed to assess the heat that is absorbed in the top 2000 meters of the ocean. This is where most of the heat accumulates. Most of the data comes from 3,800 free-drifting Argo floats that transmit it to satellites.

Satellite studies by Anna Folgmann, a researcher at the University of Leeds reveal that a massive iceberg that broke off from Antarctica in 2017 and slowly drifted towards the sub-Antarctic Island of South Georgia last year, has released an astonishing 152 billion tons of fresh water into the South Atlantic as it disintegrated. "This is a huge amount of fresh water melt; what we want to learn is whether it had a positive or negative impact on the ecosystem around South Georgia, "she said.

Antarctic Greening: Scientist have documented what they call a "striking expansion" of Antarctica's two native flowering plants, driven mostly by a rapidly warming climate. Writing in the journal Current Biology, they documented how Antarctic hair grass and Antarctica pearlwort have expanded on Signy Island, off the Antarctic Peninsula, as the average summer air temperature increased by almost 3.6 degrees Fahrenheit since 1960. The report noted that this spread will alter the local ecology. The National Oceanic and Atmosphere Administration (NOAA) climate scientist say that during 2020, the planet experienced the seventh warmest January since reliable records began in 1880.

Climate change is what people talk about but what does it mean? Changes to what? We must go beyond that simple expression and consider what is really important; the changes in the biosphere and how such changes will impact humankind. The biosphere includes the atmosphere, the ocean's hydrosphere and the earth's lithosphere. This is where the essentials for life on the planet exist and it is the changes within these systems that will determine the future of the planet that humans and all living things utilize.

What have been the changes in the atmosphere, the earth and the oceans that have already occurred? We know that our atmosphere has been saturated with carbon that is creating excessive heat around the world. We know that the magnitude of the rate of warming over the last 150 years far surpasses the magnitude and rate of changes over the last 24,000 years. We know this heat has warmed the oceans; damaged the coral; increased evaporative moisture, started the melting of the ice caps and glaciers; caused extensive severe storm systems and flooding; created an unbalance in the oceans currents and raised its elevation. We are living today with these known changes and they will continue.

All of these changes are established and none of them are exclusively local, they are universal worldwide around the globe. The big question is how are these changes affecting mankind and the other living creatures he shares the planet with now and in the future? It is clear that humankind is the most influential and destructive force on earth. Humanity is on trial, with little time left to fix things before the verdict is in and the planet imposes its most severe penalty. This is not the message that political leaders, policy makers or friends and neighbors want to hear. Unfortunately, it is the ultimate inconvenient future.

Today it is disturbing that many people will not accept or even hear of, or observe that their world is changing around them; that their way of life may be on the verge of collapse. No, not tomorrow, next year or hopefully not in five hundred or even in a thousand years;

but man's actions have started the clock ticking toward a wake-up call and the hangover will be severe. Preparing for long-range disaster or even thinking about it is hard to do because it seems like a waste of energy and resources. We are natural optimist even after experiencing a tragic event. "Oh, it won't happen again or it won't be as bad or affect me." We slack off and ignore the present and emerging climate changes because it is something we can put off; we refuse to accept it or don't understand what's happening or think it won't happen to us. It will.

Today, encouragement, information and teaching from a leadership that is informed and acts without bias is an absolute necessity if the public is to grasp the importance of what's happening to their lives, the lives of their children and all future generations. Can we find such leadership and what if we don't?

 Carbon dioxide concentrations in the atmosphere from human activities has increased by 47 percent, driven by economic and population growth since the start of the Industrial Revolution. Before the first industrial industries began burning large amounts of fossil fuels in the late 1800s, it was estimated that the atmosphere contained around 280 parts per million (ppm) of carbon dioxide as measured in ice cores obtained in Antarctica. In 1958 when measurements were begun on Mount Mauna Loa in Hawaii that level was 310 ppm. Today the increase stands at 419ppm and is growing. Increases of water vapor, methane, nitrous oxide and chlorofluorocarbons are also contributing to the greenhouse effect. As the planet warms the weather patterns have become altered; sea levels have risen due to the melting of Earths glaciers and ice sheets and heat expansion; Ocean and land heat waves have become more frequent, lasting longer and are more intense. Human activities are changing the natural protective greenhouse.

Many climate skeptics still don't believe climate change is caused by humans, despite overwhelming scientific evidence to support that conclusion. And there are those that don't even acknowledge that the change is happening at all. Some deniers claim these fluctuations in the Earth's atmosphere are being caused by the change in the Earth's orbit, the oceans circulation variations, El Nino, land use changes, air pollution and smog. It has been proven that none of these produce the actual changes that are occurring.

Man must obtain his knowledge and chose his actions by the process of thinking which nature will not force him to perform. Man has the power to act as his own destroyer and that is the way he has acted throughout history. Ann Rand

There must be a change in the political cultures that redirects efforts toward action and not denials. We no longer need focus groups who have their own agendas of maintaining the status-quo or outright opposition to what needs to be accomplished. It is easy when it comes to climate change to take action as the public demands. Make a lot of grand statements as to what needs to be done about heat waves, flooding etc. then forget about them when people complain about high gas prices. In other words, talk a lot but back off when it comes down to taking difficult constructive action. It is easier to deny a harsh reality then it is to fix it. This has been the practice in the past but hopefully its seems this is changing. People are at last becoming aware of what's happening with the climate and are beginning to demand positive change. It's too late to pick the low hanging fruit, its already spoiled and fallen to the ground. We need to reach higher for the difficult stuff like seriously reducing carbon emissions by limiting the use of coal, oil and gas and placing a carbon tax on carbon releases. Disaster preparation will always be inconvenient and challenging but necessary.

If we are to see meaningful climate legislation pass in Congress, it will require a massive turnout in the elections by voters who value and prioritize a livable climate. Will that happen? Are the American people going to wake up and realize how important their decisions are for themselves their children and future generations? We need a Congress that is willing to pass a major climate bill that incentivizes renewable energy, places a cost on carbon emissions and disincentivizes fossil fuel infrastructure and extraction. Half measures recently attempted and passed, although politically calming, will not slow down or stop the earth's burning. The use of oil, gas and coal must be drastically reduced and can only be done by strong no nonsense permanent government action. The recent Congressional activity on climate change is a good start but it depends on too many small achievements by the public, the government and corporations. It is a smorgasbord of proposed measures that will have questionable results; but it feels good and makes effective news headlines and spends a lot of money.

Ignorance is a virus and once it starts spreading it can only be cured by truthful information and intelligent thought. Byrum

It appears that we are again trying to buy ourselves out of the problem by massaging public opinion and spending billions of dollars where a simple direct act of Congress to tax carbon and cut the use of coal would accomplish much more, with quicker results and would be a whole lot cheaper. We are playing gutless politics with the future of the earth and its people and there will be no winners. Compelling actions must be taken now if the United States is to be in a position to lead and meet its obligations to the rest of the world by cutting its carbon emissions in half by 2030. There will be little hope for a livable secure future for our children and their children if we pass up this opportunity.

People living today must come to realize that global temperatures are rising at an unsustainable rate. It is the largest rise in human history and the most rapid rise in earth's 4.5-billion-year history. If the amount of emissions that we are currently sending into the atmosphere are not drastically curbed, the increasing earths temperatures will spin out of any control in just a few decades, and we will lose any possibility to influence that escalation; causing massive changes to our environment. Think a minute, have these changes already started and have we become immune to their effect? Have we started down the road to the earths sixth mass extinction as some scientist predict?

"Wisdom cries out in the streets; in the market she raises her voice: How long O simple ones, will you love being simple? How long will scoffers delight in their scoffers and fools hate knowledge?" Proverbs 1:20

Today the principle of "cause and effect" has been reversed. Normally "cause" is the unknown and sometimes it is very difficult to determine why or how an event is occurring. Ask any doctor, they will deal with this question on a daily basis. That is certainly not the case with climate change. We know and have known for a very long time what has been happening to our world and why, and finally people are beginning to understand that our planet is changing; they are seeing it and many, unfortunately are experiencing the consequences themselves.

The big question confronting us today is "effect." What will be the outcomes of these changes now and in the future? How hot will it get and where? How high and how fast will the oceans rise and how much heating will they experience? Where will we be able to live, will we have to migrate because of warmer weather or rising seas? How will our food supplies and its production be affected, will there be enough food to available for the worlds people? Will drought and fresh water shortages continue and increase? Will weather systems; floods, hurricanes, tornados increase in numbers and intensity? Will we be able to deal with the discovery of new viruses and bacteria? Will the Artic tundra continue to melt releasing vast quantities of earth heating methane, how much? Will we be able to control these changing events and maintain a livable viable Earth? These are but a few of the unknown effects that man will be challenged with, certainly, there will be others.

There is a practice currently being advocated by the Worlds Governments and major corporations called "Net Zero." It is an attempt to achieve an equalization of the amount of greenhouse gases we place in the atmosphere with the amount we remove, a balance achieving zero additions. But how is the balance to be achieved, the gas won't actually be removed, just relocated? We will still be burning carbon and hoping it is being sequestered in other locations around the world by nature such as trees and forest. How is that to be measured? We don't dare count on the oceans to continue to absorb increasingly large amounts of carbon, they have already reached a limit that is not sustainable as witnessed by their increased heating.

To date we do not have the technology to actually remove carbon from the atmosphere in the amounts that would make a difference. The target date to achieve a net zero balance has been set at 2030, just in eight short years. It's a farce, it will never happen, it's a ploy to delay immediate corrective action in order to even come close to achieving such reductions. According to the scientific community, emissions would have to be reduced by 45% worldwide. But projections show that given the current emission rate there will be a net increase of over 15% during that time period. So, the target maximum temperature increases of 1.5C in 2030 will not be achieved. We will blow past that target to much higher temperatures creating a perilous future of conditions beyond our control. Even though it is a small step in the right direction the campaign for electric cars will not help much. Robert Stavins, a professor of energy and economic development at Harvard's Kennedy School said, "Individual action on climate change is not going to be sufficient in 10 or even 20 years to make a significant difference. We need policies for industry and governments to move forward now in a carbon friendly direction." If we could shut down the use of coal worldwide then there might be a chance for significant reductions, as it stands right now that's a "fat chance."

But let's just suppose that the world is able to reduce the greenhouse gas emissions by the target amount of 45% by the year 2030. This does not mean that we will be reducing the amount of existing carbon in the atmosphere, it just means that we will be burning 45% less. But what happens to the remaining 55% that the world is still burning? It does not seem logical that we will ever be able to achieve "carbon neutral," where we are able to balance the amount being burned with the amount contained and stored by nature. The percentage of the residual gasses that cannot be sequestered by nature in the form of trees or the oceans; and remember the oceans are already heating and saturated with CO2, has to go somewhere and that somewhere is the atmosphere where it will join and increase the quantity and heating effect of the existing carbon dioxide. Today, and in the foreseeable future, we will not have the technical ability to capture and remove but a small portion of the existing huge amounts atmospheric carbon that is being released.

So, what have we accomplished? We have certainly slowed down the amount of carbon we have been releasing and in so doing reduced the rate of future warming; but we haven't stopped it. Additional greenhouse gases will still be flowing into the atmosphere and until we have the ability to remove them; not slow them down, the earth will continue to heat up although at a slower pace. The continuing increase in warming can only be controlled by reaching zero emissions and that will never happen. So, what can reverse it? There seems to be two chances, neither of them good. Mother Nature can stop it as she has done many times in her long past with catastrophic events such as another ice age caused by massive volcanic eruptions that block out the sun for an extended period, or we will be able to develop the science that is able to remove huge amounts of the existing carbon rapidly. But we had better hurry.

We use the years 2030, 2050 and the end of the century as arbitrary times to achieve emission reduction goals to maintain reasonable temperature increases over those time lines. But they are just arbitrary target dates that set no guidelines to even consider what will occur later. Let's assume that we are able to hold the temperature increase to 2 degrees Celsius by 2050 or even by the century's end. What happens then, have we stabilized the atmospheres temperature? No, the earth will still continue to warm, it won't stop warming at the end of the century, we will still be releasing carbon into the atmosphere that will merge with the destructive amount that is already there. Hopefully, it may be considerably less than the amount of today's emissions with the rate of heating and melting reduced, but it will not stop. So, what will the global temperatures be in 2150, 2200, or beyond, will we still have a livable world? For those who think that it is a long time in the future, consider your children's future children and grandchildren, they will be living it.

The extreme weather events the world is experiencing today will continue to increase in frequency and intensity as we keep warming the atmosphere by adding carbon pollution by burning fossil fuels of coal, oil and natural gas. It's not only the temperatures that must be considered it is the projected damaging effects on all of Earths ecosystems, the global economy and how our way of life will change.

Climate change and the increasing weather extremes it creates is changing the real estate equations. The American dreams are increasing running into weather nightmares and raising pressing questions as to where it makes sense to live. People will have to make decisions about their lives that they never would have expected before the advent of climate change. After having experienced severe storms, flooding, and wildfires or extreme killing heat what do you do, will you have to move? Can you, and what if you can't? Will you be able to endure and afford another flood or fire disaster; another hurricane or try to cope with the unbearable continuing warming? This is the tragic human side of our changing world and there are no simple answers.

The hotter and dryer weather caused by the changing climate has nearly doubled the amount of forest that has burned around the world in the past twenty years. Last year, 2021 was the worst burning event ever with an area of 34,750 square miles of forest consumed. More than half of the destruction occurred in Russia. "It is astonishing how much loss of forest cover has occurred over such a short time," said James McCarthy, an analyst with Global Forest Watch. In the Amazon, the largest rain forest in the world and an area that sequesters huge amounts of carbon dioxide, illegal deforestation has led to thousands of fires. As these fires burn and the trees are removed that carbon is released.

The Brazilian National Space Institute said satellites have detected, mostly illegal, 33,000 fires burning in the Amazon in the first half of 2022. The majority of these fires are man caused fires set to burn and clear the forest. The number of fires this year has increased more than 20% over 2021 and the Amazon has had its worst month in the history of forest devastation. The forest may grow back in a hundred years but their burning releases huge amounts of carbon into the atmosphere increasing the Earth's roasting. The obliteration of the forest for new agriculture; soybean production that is sold to China to raise chickens and pigs, and cattle ranching continues at an alarming rate. By this destruction of the Amazon, Brazil has become the largest soybean producer in the world. It may soon happen that the Amazon Forest will become an area that releases more carbon than it stores. A loss of a forest to raise chickens!

A research paper published in the journal Environmental Pollution found that the California wildfires in 2020 caused twice the amount of greenhouse gas emissions then the state successfully reduced between 2003 and 2019. Eighteen of the largest wildfires in California's history have occurred since 2000. The eight largest fires have all happened since 2017, five of them occurring in 2020. During 2020 more than 9000 wildfires took place in the state sending smoke all the way to the East Coast. Economic losses exceeded $19 billion dollars while a total of 4.3 million acres burned.

The 2020 wildfire season wiped out 16 years of positive progress that California had made on climate change through efforts such as fossil fuel replacements and clean energy. In other words, the wildfires that have scorched the West in recent years are not necessarily the consequences of climate change as they are an increasingly sizable driver of the problem; producing tons of carbon dioxide for a warmer climate. Essentially, the efforts and all the hard work to reduce emissions over almost two decades was swept aside by the burning produced by a single year of record-breaking wildfires where more carbon was released than the amount reduced over that time period.

There are steps that can be taken that can lessen the impact of these changes. National legislation must be passed that puts a price on carbon emissions to hasten the transition to a clean energy economy and encourage the development of solar and other forms of renewable energy. There are bills proposed in Congress to do that but unfortunately too many political hurdles have been put in place by powerful political and corporate interest to get meaningful legislation established. Even the Supreme Court, at the very time we are experiencing the devastating destructive consequences of our past inaction, has restrained the Environmental Protection Agency's efforts to limit carbon dioxide emissions from coal fired power plants. Democratic Senator Joe Manchin, from West Virginia, a coal producing state, has been the cog in the wheel to get any meaningful climate legislation passed even as his state is experiencing major weather events and severe flooding. As of the current date he has defied his party and has reaped millions of dollars from the coal industry for his support of that industry, and he has refused to vote to fund any climate change legislation.

Scientific facts and even the obvious occurring severe weather events of today have failed to impress upon the public a need for immediate action. The "financial bottom-line," political pressure and obstructive politics rule and if that remains the case, informed and concerned policy makers trying to implement common sense climate changes have their hands tied. We are living in perilous times when politicians and financial interests control the future health of our planet and its people by ignoring scientific facts and by refusing to take decisive action. How long must we wait, and will it be too late, before policy changes can be implemented that seriously and forcefully attack the problem?

"Man can hardly recognize the devils of his own creation. He has lost the capacity to foresee and forestall, he will end up destroying the earth." Albert Schweizer

Will we revert again to another "do nothing" administration because the people do not understand the importance of their vote in changing the direction the country and the world as we head to a much hotter and wet future? Many have not believed the scientists that have been warning them for years about what is occurring today with the severe weather and the world's increased heating. Some unfortunately, may have already been made "Christians" as they have experienced wildfires, severe storms, flooding, and extreme heat firsthand. That's a hard, painful, and costly way to learn but everyone had better start learning that climate change is real, hopefully not the hard way.

In an early sign of how the incoming Republican House majority, after the 2022 mid-term elections, will treat climate change, *Bloomberg News* reported that the current House Select Committee on Climate Change will be eliminated when the Republicans take over. Rep. Garret Graves, the Louisiana Republican who was his party's highest-ranking member of the committee, said Republicans will focus their energy agenda on "increasing U.S. fossil fuel production and exploration." Does that answer the question about the "do nothings"?

This will confirm the concern of climate activists of what would happen if there was an administration change. "I think the fact that every single Republican voted against the Inflation Reduction Act, that contained provisions that addressed the climate, tells you everything you need to know about where Republicans stand when it comes to climate change," said Senior Vice President of Government affairs at the League of Conservation Voters.

International Finance, a policy just approved at the 2022 climate congress in Egypt, and so far, the most important agreement passed by the 193 countries in attendance is in serious jeopardy and could result in the failure of the conference to pass any meaningful climate agreement. The U.S. pledged, as part of the agreement, to contribute $11. 4 billion dollars by 2024 for aid to developing nations that are trying to reduce their emissions and adapt to climate change. But this year after the administration requested $3.1 billion for climate assistance, the Democratic Congress appropriated only one billion. Experts say it is extremely unlikely that the Republicans will agree to increase that amount now that they have control of the House of Representatives.

So again, has the U.S. made a climate promise to the world that it may be impossible to meet and how are we to expect other nations to sustain their obligations? Most climate activist, scientist and other experts including their Democratic allies in congress, disagree with the Republicans who are advocating that fossil fuel production should be increased and that dispute will make it almost impossible for any meaningful climate legislation in the next congress. So here we go again, another two years lost in the efforts to control the Earths warming and all because of nonsensical political B.S.

Nothing today is more important; not abortion, gun control, the economy, immigration, inflation: Nothing! People must be made to realize that how they vote and who they vote for will affect not only them but their children, their grandchildren and all future generations. The die is cast, we must slow down this run-away truck that is taking the world to oblivion. That can only be accomplished by knowledgeable and forceful leadership. We had four years of an administration that made absolutely zero effort to control climate change, even to the point of calling it a "hoax" and denying it even existed. We cannot afford to waste any more time, we must act, and it is the people through their vote that must elect informed concerned and forceful leadership that will install meaningful climate policies if we are to have any chance to control what is happening to our world and our children's and their children's future.

Unfortunately, today we do not have the technology to eliminate or reverse the atmospheric damage that has already been done by removing carbon from the atmosphere; it is essentially permanent and increasing. All we can do is to try and slow the rate of current emissions. The heat, drought, flooding, increased fire occurrence and intensity and increased damaging storms; all of these are now fixed in our future. They are not going away. They will verry in frequency and strength over the coming years but the destructive effects will remain and increase.

We must learn to live with the current conditions and try to minimize the addition and effects of any future emissions. But we better get serious, we are running out of time. Carbon dioxide, the major greenhouse gas, will stay in the atmosphere for hundreds of years. Yes, there are events that could cool the earth and they have happened many times long before man appeared. A massive volcanic eruption like the one that occurred in Yellowstone National Park or a major meteorite strike would cool the world down by rapidly filling the atmosphere around the globe with millions of tons of debris, blocking out the sun; instant cooling. But be careful what you wish for.

From the Journal Advancing Earth and Space Science

The last remaining North American ice sheet is projected to disappear within the next 300-500 years. This is the last remaining ice that once covered North America. The Barnes Ice Cap located on Baffin Island in the Canadian Arctic is all that is left of the Laurentide ice that covered millions of square miles of North America roughly 2.5 million years ago. This remaining ice has been stable for about 2000 years until the effects of the recent warming began its current melting.

A new study done by Gwenn Flowers from the Simon Frazer University determined that the ice cap is currently melting at all elevations at a rate of about 3 feet per year and will be gone completely in three to five centuries regardless of any additional carbon dioxide emissions. They found it extremely rare for this ice cap to melt completely. Rock samples taken from the ice cap and examined by carbon dating determined that it shrunk to its current size and disappeared completely no more than twice in the past 2.5 million years. "The geological data is pretty clear that the cap almost never disappears in interglacial times," he said.

"The new study is yet another demonstration that humans are pushing Earth's systems beyond conditions that have existed over millions of years," said Richard Alley a glaciologist and climate scientist at Pennsylvania State University. "We have very high scientific confidence that almost all the glaciers and ice caps on

Earth are shrinking because of warming caused by the emissions of greenhouse gases," he stated.

For more than 30 years David Harwood, professor in the Department of Earth and Atmospheric Sciences at the University of Nebraska- Lincoln, has been a leading scientist investigating how global temperature change has affected the Antarctic ecosystems and the stability of the large southern ice sheets over millions of years. His studies have provided a history of the ice sheets obtained by many drilled ice cores that reveal when the amount of atmospheric carbon was the same as it is today, and what it may possibility be at the end of this century. As an example, the last time carbon levels were what they are today, 419ppm, was 4 million years ago and the result of the warming at that time when the West Antarctic Ice Sheet was melting, the sea levels were several meters higher than they are today. Currently, the Antarctic glaciers are melting in specific regions of West Antarctic at a rate that can't be stopped. It is estimated that the huge Thwaites Glacier is as at a point where it could collapse into the ocean within a decade. The professor said: "This is not a scientific problem anymore; the science is clear. What we know is scary and we are overdue for deliberate action to reverse course."

Over the last 70 years the average number of frost-free days in Anchorage, Alaska has increased by 17 days. That is significant in terms of shorter winters and longer growing seasons. The temperatures in Alaska and the rest of the northern latitudes are rising at a faster rate than those of the lower states. Instead of the normal reliable 50-degree summers in the 1980s, Alaska now has a balmy average of 60 degrees. Alaska serves as a bellwether for the rest of the world and it could be in the future, the last bastion of relief in the inferno of climate change. Heat waves will still damage the lands, its people and the environment and will have devastating effects on the salmon runs and their populations by decreasing stream flows and warming the aquatic ecosystems. The rising temperatures will not be as severe as other places in the country and Alaska, may in the future, serve as mainstay for a sustainable place to live.

The good thing about science, "Its true whether you believe it or not, that's why it works." — Neil de Grease Tyson

The Artic Report Card. This report has been published yearly since 2006. It is a scientific review of reliable and concise environmental information pertaining to the Artic environmental ecosystems. "This year's report continues to show how the impacts of human caused climate change are propelling the Artic region into an entirely different environment than it was just a few decades ago, "said NOAA Administrator Rick Spinrad. "The trends are alarming and undeniable. We face a decisive moment and must take bold action now to confront the climate crisis."

Some of the findings of the 2022 report:

The average surface temperatures over the Artic from October 2020 to September 2021 was the warmest on record. This is the eight consecutive year since 2014 that surface air temperatures were at least 1 degree Celsius above the long-term average.

1. August sea temperatures warming trends for 1982-2021 show significant warming for the Artic Ocean.

2. Snow cover in the Eurasian Artic in June 2021 was the 3rd lowest amount since records began in 1967. In the North American Artic snow cover has been below average for 15 consecutive years.

3. The Artic continues to warm twice as fast as the rest of the globe.

4. The Greenland Ice Sheet experienced three extreme melt episodes in late July and August. On August 14, 2021 it rained at the 10,500-foot Summer Station for the first time ever.

5. Substantial decline in Arctic Sea ice since 1979 is one the clearest indications of climate change. The summer of 2021 saw the second-lowest amount of older multi-year ice since 1985 and the winter sea ice volume in April 2021 was the lowest since records began in 2010.

6. Ocean acidification in the Artic Ocean is occurring faster than the other global oceans. A growing body of research indicates that acidification in the Artic Ocean could affect Artic ecosystems, including influences on algae, zooplankton, and fish.

7. Retreating glaciers and thawing permafrost are causing local and regional-scale hazards that threaten lives and livelihoods, infrastructure, future development, and national security.

8. Beavers are colonizing the Artic tundra of western Alaska for the first time, changing tundra ecosystems and degrading permafrost.

Beavers have found their way to the far north. They are developing homes and colonies throughout the tundra of Alaska and Canada and their numbers are increasing. Why does this matter? How they are affecting the Artic ecosystems is not yet entirely clear, but climate scientists are concerned. Researchers have observed that the dams bevers build can accelerate a warmer climate. "Beavers really alter ecosystems" says Thomas Jung, a senior wildlife biologist for Canada's Yukon government. Bever dams affect everything from the height of water tables to the kinds of shrubs and trees that grow. Bevers relay on woody plants for food and materials to build their dams and lodges and the rapid warming of the Artic has made the Arctic more hospitable for their growth. Early snowmelt, thawing permafrost and longer growing seasons have increased the growth of plants like alder and willow that beaver thrive on.

In the 1950s there were no beavers in all of Alaska. But a recent study by Ken Tape, an ecologist at the University of Alaska, Fairbanks scanned satellite images of nearly every stream and lake in the Alaskan tundra and observed 11,377 bever ponds. He found that the number of ponds doubled between 2003 and 2017. These ponds are thawing permafrost and effecting climate change. Bever dams create warm areas because the water in the ponds is deeper and doesn't freeze all the way to the bottom in winter. The warm ponds melt the permafrost; the ground then releases the stored carbon in the form of greenhouse gases; carbon dioxide and methane, that contributes to atmospheric warming.

In Earth's history when the amount of CO_2 in the atmosphere was about 500ppm large areas of forest grew in the Arctic. During these levels of high greenhouse gases there were few places on earth where the climate was too cold for trees to grow. In the ancient past, as now, the polar regions warmed faster than the rest of the planet. And since the human contribution of fossil fuels has now raised CO_2 levels, the intensified warming of the tundra appears unavoidable as does the spread of animals and plants that need those conditions to survive.

Swiss scientists declare that the country has lost half of its glacier volume since 1931. The researchers say they "reconstructed" the topography of Swiss glaciers as they appeared in 1931, and then followed how they slowly disappeared over nearly 100 years of melting. Their study included 21,700 photos of the glaciers taken between 1916 and 1947. As an example, the Fiescher Glacier was a massive sea of ice in 1928 but now is only a few tiny specks of white.

Glacier National Park: During the 19th century Glacier National Park had 146 small mountain glaciers. By 2005 the number of glaciers in the park had shrunk to just 32 and by 2015, the latest area survey, only 26 remained. It is predicted that by 2030 the park will need a new name when there are no longer any glaciers. Glaciers perform many ecological functions. They store water in the summer and release it slowly that cools streams and rivers protecting fish and their ecosystems. They provide cooling regions for large animals such as mountain goats and grizzly bears during the hot summer months and in many regions of the world they supply drinking water to millions of people. Glaciers are important in that they provide sensitive insights as to how the planet is responding to climate warming.

Since 1992 the sea Arctic ice cover has been reduced by 40%. Mountain glacier fields around the globe have melted 6 trillion tons of ice. Greenland alone has lost 4 trillion tons. The International Panel on Climate Change estimates that under current glacier melting rates, long term ocean levels could rise 35 feet or higher. Dan Schrag, the director at the Harvard University Center for the Environment said "Even if we could become carbon neutral tomorrow, the climate will keep changing for thousands of years, the ice sheets will keep melting and the seas will continue to rise."

In 1960s humans released about 9 billion tons of CO2 a year, now we are releasing about 35 billion metric tons annually. This rate of increase is 100 times greater than any previous period in earths geological history. There are some nations and many people who have adopted the shear inertia of hopelessness in the face of global warming. Rather than trying to control the warming they have found it much simpler to find excuses for continuing with business as usual. This is an attitude in which certain vested interest in some of the wealthy nations have clearly succumbed. Some countries actually welcome a warmer planet. Russia has a lot to gain with an ice-free Arctic including an increase in shipping, oil development in the northern oceans and vast new land areas that would be available for agriculture and development.

Climate Starvation: Hunger brought on by climate change appears to be behind the deaths of hundreds of small Korora penguins that have washed up on the northern shore of New Zealand since last May. Also known as little penguins, they were found through autopsies to have starved to death and were not victims of disease or toxins. Warmer waters now force the small fish they eat to swim deeper to reach cooler temperatures. Since the penguins can only dive to depths of 65 to 100 feet, they are increasingly unable to reach their prey.

Life in some places of the planet is readily reaching the point where it will be too hot for the species that live there to survive according to a report just released by the Working Group for the Intergovernmental Panel on Climate Change (IPCC). It addresses scientific facts documenting the change on ecosystems and society worldwide. The report prepared by 270 top scientists from 67 countries states, "human -induced climate change, including more frequent and intense extreme events has caused widespread damage to nature and people beyond any natural climate variability."

Hoesung Lee, the head of the IPCC said the report is a "dire warning about the consequences of inaction. It shows that climate change is a grave and mounting threat to our well-being." Delays in taking action caused by misinformation and inaction addressing climate change, have made the need to act even more urgent. This has led to increasing uncertainty and slowed the response to the risks. The co-chair of the report states: "Climate change is a threat to human well-being and the health of the planet. Any further delay in concerted global action will miss a brief and rapidly closing window to secure a livable future."

"The latest State of Global Climate report is a chronicle of climate chaos," said the United Nations Secretary-General in response to the reports recent release. "As the World Meteorological Organization (WMO) account shows so clearly, change is happening today with catastrophic speed, devastating lives and livelihoods on every continent. Glacier melts are jeopardizing water security for whole continents. We must answer the planets distress with action now." The world is falling far short of its climate change goals with "no credible pathway in place to keep the global average temperature increase below 1.5 degrees Casius," the UN just reported. Emissions must be cut by as much as 45 % world- wide by 2030 to avoid tragic consequences from more intense and longer heat waves and drought. "Loss and damage from the climate emergency are getting worse by the day" said the UN secretary.

"It is an enormous challenge but an absolute necessity. We need to transform our industrial energy and transport systems along with our whole way of life." Said the leader of the World Meteorological Organization (WMO). These drastic reductions are needed to contain the trend in rising temperatures and to prevent a 6-degree Fahrenheit increase by the end of the century. All three of the main greenhouse gases, methane, CO2 and nitrous oxide reached new highs in the atmosphere last year, the WMO reported. Methane saw the largest rise since measurements began 40 years ago and CO2 levels were higher than average over the past decade. "We are headed in the wrong direction," said the WMO. Progress on emission reductions has been nowhere near the scale and pace required. The report looked at 40 indicators, accounting for about 85% of all greenhouse gas emissions globally and could not find any on track to meet the 2030 targets.

However, there is some good news and hope for a turnaround in the direction we are headed although if it occurs it will be in the very distant future. In California, scientist at the Lawrence Livermore National Laboratory for the first time have managed a small advance in the development of fusion energy technology. Although unfortunately, there are huge roadblocks to be over come before fusion energy can power our cities. The single most important drawback is the fact that the experiment did not create more energy than it used in the process. What was announced was that an amount of 2.5 megajoules of energy was generated by the laser and 3.15 megajoules were produced. But Mark Herrmann, Livermore Laboratory program director acknowledge this leaves out the huge amount of electricity used to power the laser. The experiment drew about 300 megajoules from the grid so in reality the results produced only 1% of the energy used to produce it.

"Given the massive technological challenge that will be required to actually utilize this technology in a reliable and affordable way of generating electricity for the electric grid, I think it has little or no relevance to the climate change crisis," said Edwin Lyman, a physicist who serves as director of nuclear power safety with the Union of Concerned Scientist, "The climate crisis has to be addressed now and technology that's going to take many decades of possible development is not going to play a role."

There are many unanswered questions about the practicality of using fusion to generate electricity for this country and the world. Can existing power plants be refitted for that purpose, or will new plants and transmission systems and grid upgrades be required? Then there are the materials and components, which will be needed to withstand exceptionally high temperatures and radiation levels. While fusion doesn't produce long lasting levels of radiation waste in a way fission does; it relies on tritium, a radioactive isotope of hydrogen. Tritium exists in only trace amounts in the atmosphere, it can be produced in a nuclear reactor but that is not a practical and affordable process. The future fusion power plants will have to breed their own tritium and at present there are no systems yet designed that can do that. "The energy that is produced in the fusion process is mostly heat so to convert that into electricity it must be changed to steam for steam generators that are only about 40% efficient." Said Michael Mann an environmental scientist at the University of Pennsylvania. "While it is important to continue fusion energy research, today is not some sort of panacea with regard to the climate crisis. The reality is that fusion energy will not be a viable alternative for many decades."

Alain Brizard, a theoretical fusion expert at Saint Michaels College in Vermont said," it may be possible in the next 20 years to have a commercial operating nuclear fusion reactor." How long it will take for worldwide application and installation of new reactors is unknown; and what will be the cost? Let's hope we are not too late. The downside to this breakthrough is that it is not going to be the answer to climate change and doesn't mean that efforts to shift to carbon neutral energy production now can be stopped. "There is no chance that this will be commercially available by 2030 when the world must reduce emissions severely if we hope to keep temperature rise at a reasonable safe level," said Michael O'Boyle, director of electric policy at Energy Innovation.

We are like the little boy who stands at the face of a dam with his finger placed to try to stop the small leak while the water floods over the top.

Tripping Points

What are they? As referenced to climate change, they are the locations and time when events that are occurring to planetary systems reach a point of continuing alteration that is irreversible. This is a point when natural events have become so out of balance that any continuing destructive causes cannot be overcome or slowed down as they continue to occur and increase. Tipping points can also be described as a moment of critical change in human history.

Always in the past the Arctic was known for maintaining year-round ice and snow but just in the last decade it has begun to change into wetlands and open ocean. The last intact ice shelf in the Canadian Arctic fell into the sea in July of 2020. Since it was first studied in 1902 this ice sheet had lost 43% of its mass. Canadians Ellesmere Island ice caps were also lost in the summer of 2020 when they melted completely. This glacier melt, thawing permafrost and wetland expansion create new landscapes that change ecosystems as well as alternating the global atmosphere and ocean circulation. These are examples of tipping points occurring today.

The Arctic yearly lows of sea ice formation and an alarming increase in the permafrost thaw occurring in today's warming climate signals that a tipping point has already been reached. We may have already lost the frozen Arctic. This is a difficult thing to accept but the evidence is irrefutable. There is very little time left to alter the changes occurring in the earth's ecosystems to halt their climate driven collapse. To protect the earths diversity and stability we must accept the fact that the climate is already rapidly permanently changing the planet. Without decisive leadership and firm action to reverse the proliferation of greenhouse gases these changes will continue at the worlds and its people's peril.

An article in USA Today cites a recent United Nations panel on climate change that outlines five critical events that scientist project could reach tipping points in our children's and their children's lifetimes. In a recent speech the U.N. Secretary – General said "the world is perilously close to tipping points that could lead to cascading and irreversible consequences."

1. Savanna Erases Amazon Rainforest
2. Death of Coral Reefs
3. Ice Sheet and Glacier Melting
4. Atlantic Current Circulation Slows or Stops
5. The Snow Forest Disappears

Savanna Erases Amazon Rainforest

The Amazon is a 2.5 million square mile rainforest that is critical to controlling the amount of CO2 that is admitted into the atmosphere. Its millions of trees collect and hold vast amounts of carbon; it is home to 10% of the world's species. Rising temperatures, increasing drought, fires, and man's influence are changing it rapidly from a vast lush rainforest to a dry and arid savannah. Because of increasing drought large and more frequent wildfires are occurring destroying large areas of trees that don't grow back turning the area into grasslands. This destroys the carbon sequestering benefit of the trees and as they burn, adds large amounts of CO2 back into an ever-increasing warming world. The illegal logging and burning to clear areas for farming and cattle grazing add to its rapid destruction. It has been reported by the Associated Press that deforestation of the Amazon during the first half of 2022 broke all records for any six-month period. Satellite images taken between January and June show that 1,500 square miles of forest had been destroyed by man's impact in just six months. This amounts to an area that is four time the size of New York City.

Death of Coral Reefs

Coral reefs are essential to life and health of the ocean. They provide food and shelter and are home for approximately a quarter of all marine species. A large percentage of all the fish caught in developing countries occur on or around the reefs. Coral are not plants but living organisms that grow on algae living in their tissue. They live in shallow waters and need sunlight to thrive but they can only survive when water temperatures are within a narrow range. When the seawater becomes too warm, they are stressed and are bleached white and lose their color and sustaining algae. Bleaching isn't as simple as going from a living coral to a bleached white. After they expel the algae due to overheating, they turn fluorescent pinks, blues and yellows as they produce chemicals to try to protect themselves. This process of coloring doesn't last long, in a matter of days they lose this color and bleach white. If the warming continues, they die. If they manage to survive it may take ten years for them to fully recover if their balanced water conditions improve.

Australia's Great Barrier Reef accounts for about 10% of the world's coral reef ecosystems. Its network of 2500 individual reefs cover over 134,000 square miles. Australia's government scientist reported in May of 2022 that more than 90% of the Great Barrier Reef surveyed in the latest year suffered bleaching, it was the fourth such mass bleaching in the last seven years. Bleaching is caused by warming sea temperatures due to global warming. The bleaching in 2016, 2017 and 2020 damaged two-thirds of the coral.

A report released in 2021 shows a decrease of 15% of the planet's reefs since 2009 because of global warming. They are literally being cooked to death. As the oceans continue to warm the destructive effect on these living organisms will continue. This will have a domino impact on all marine systems and humans.

"Unfortunately, the stark reality is that there is no future safe limit of global warming for coral reefs." says lead author of a research study, Adele Dixon of the University of Leeds School of Biology. Even warning to no more than 1.5 degrees centigrade is too much warming for this ecosystem on the front line of climate change. The world has already warmed by 1.1C to date and the effect on the coral has been severe. A warming to 1.5C or greater will be disastrous. A report by the U.N.'s IPCC climate science panel said global temperatures could reach 1.5C as soon as 2030. The U.N. report predicted that 70 to 90 percent of the coral reefs could be lost at 1.5C and 99% if the temperature rose another half of a degree. Coral reefs are the canary in the oceans coal mine. Human societies in the past have collapsed other ecosystems through overfishing, over hunting and development. This is what can happen to ecosystems as they reach their survival limits.

Author Terry Hughes, from the ARC Center of Excellence for Coral Reef Studies at James Cook University states that the frequency, intensity, and scale of climate fueled marine heat waves that cause coral bleaching are increasing. Five occurrences of mass bleaching since 1998 have turned the Great Barrier Reef into a checkerboard of reefs ranging from 2% of reefs that have escaped bleaching to 80% that have now bleached severely since 2016. Australia's Great Barrier Reef is the largest coral reef system in the world. A recent report by the Oceanic Atmospheric Administration Coral Reef Watch states that it is again in the grips of a record- breaking heat spell.

Ice Sheets and Glacier Melting.

Almost 90% of all worldwide transportation of goods occur on the ocean and all of the world's major shipping ports are at today's sea level. Today the world's largest ice sheets, Greenland and Antarctic and most of the land glaciers are melting adding vast amounts of fresh water into the oceans. Even if they don't melt entirely, it will still cause catastrophic sea level rise that will flood all of the world's ocean transportation hubs. Think what that will do to the economies of every nation on the planet. It may take from 100 to 300 years for this to occur, which seems like a long time. But not so long. When your children are grandfathers and grandmothers their children and grandchildren will have to live with these conditions. Consider that in the next 100 to 150 years many costal mega cities may have to be relocated. At what cost? It is easy to rationalized that this is far in the future but it must be understood that the die has already been cast. Today's melting pace will continue and increase as the world continues to warm. We have overheated the atmosphere and it's not going to cool down. The slow melting that is occurring today will continue and increase long into the future.

Dr. Michael Weber from the University of Bonn and his colleagues said at the conclusion of an extensive on-site study of the Antarctic Ice Sheet: "Our findings are consistent with a growing body of evidence that suggest the acceleration of the Antarctic ice-mass loss in recent decades may mark the beginning of a self-sustaining and irreversible tipping point of the ice sheet retreat and substantial global sea level rise."

Atlantic Current Circulation Stops

Research suggests that if global temperatures continue to rise the Atlantic Meridional Overturning Circulation (AMOC) could fail or seriously weaken in 50 to 250 years. This is the system that contains the Gulf Stream. Admittedly that is a very large time spread and there are large unknowns, but there is evidence that the system has experienced some weaking over the last few decades and may becoming unstable. This event is known to have occurred in past geological times. One of the unknowns is the rate at which the Greenland ice sheet will melt. As it continues to melt it will pour large amounts of fresh cold water into the Atlantic Ocean. Fresh water is lighter than salt water and does not sink and by not doing so disrupts the cycle of circulation, slowing it down. The AMOC carries warm tropical water that flows north along the coast of Europe stabilizing a warmer climate by several degrees. If it should slow down or stop Europe would be much colder. It also would strengthen hurricanes in the Atlantic and rise sea levels along the northeastern coast of North America because less water would be flowing and cooling; it would back up along the East Coast. A weaken AMOC means less water is transported from the tropics toward the North Pole which in turn leads to warmer waters in the Southern Ocean. This can destabilize ice sheets in the Antarctic, which increases global sea level rise and more melting at the edge of the Greenland ice sheet.

The Snow Forest Disappears

The cold weather forest that occurs across Canada, Alaska, Siberia, and North America are believed to contain 25 to 35 percent of all the carbon on the planet. Even a partial loss of these forest by fire, heat or insect infestation would emit vast quantities of greenhouse gases into the atmosphere adding to global heating. Rising temperatures are causing drought that is stressing the trees and drying out the eco-system, making them venerable to insects and adding fuel for larger and hotter forest fires. After the fires, grasslands and shrubs will replace the trees, and summers may become too warm for the trees that were adapted to the cold to return.

And then the question remains; how close to we have to come to the tipping points of irreversible climate change do we need to come before people's infinite capacity to calmly watch their own demise is ended?

Stephen Hawking:

Four Life Lessons from a True Genius;
By Karla Peterson, Columnist, San Diego Union Tribune

It is no secret that when Stephen Hawking died in 2018 the world lost one of its most brilliant minds. He spent most of his life solving some of the world's most complicated problems. He had a lot to say and some of it impacted everyone on the planet. Hawking made some dire predictions on how the planet was going to end and how human kind was going to destroy itself.

When America withdrew from the Paris climate agreement most scientist around the world were dismayed, they could not understand it. Stephen Hawking at that time stated:

"We are close to the tipping point where global warming becomes irreversible. Trumps action could push earth over the brink to become Venus with a temperature of 250 degrees." Hawking when speaking about science: "It seems as if we are now living in a time in which science and scientists are in danger of being held in low, and decreasing esteem. This could have serious consequences." He continued: "We have come to expect a steady increase in the standard of living. But people distrust science, because they don't understand it. Without a basic understanding of scientific principals and trust that scientists are right, bad decisions are bound to be made, everyone needs to have a basic understanding of science to make informed decisions about the future." And this was one thing that had the ability to end the world, poor decisions about genetic engineering and climate change. Stephen Hawking didn't live to see the COVID pandemic, but if he had seen it and the culture of science-denying anti-vaxxers that sprang up all over the world, he may have said something like, "I told you so."

There are over 8 billion people on earth today and it is estimated there will be between 9 billion and 11 billion by the end of the century. He stated, "Our Earth is becoming too small for us, global population is increasing at an alarming rate and we are in danger of self-destruction." While scientist say a lower population is in the best interest of the planet there are other voices with a different opinion. Pope Francis made headlines for condemning couples who decide against having children, calling them "selfish." "Sometimes they have one and quit and this is a denial of fatherhood and motherhood taking away our humanity." And therein lies the problem.

In 2016, he gave a lecture at Cambridge University, where he was very straight forward about the future saying: "I don't think we will survive another 1000 years without escaping beyond this our fragile planet." The following year he cut that time line in half to 500 years.

Some readers may question who this guy was who could make such outlandish predictions. Maybe the following story of his life will help all of us understand him better.

You don't have to be an intellectual giant to appreciate Stephen Hawking, you just have to be human. Hawking died Wednesday March 7, 2018, in his home in England at the age of 76, and the fact that he lived so many decades longer than medical science said he would, is one of the least amazing things about him. When he was diagnosed with a degenerative motor disease at the age of 21 Hawking was given just a few years to live. What we will remember most about him is not that he lived for 55 more years, but what he did with those 55 years and what we learned from them. While Hawking was given the miraculous gift of unexpected longevity the world reaped the benefits. In memory of a single man here are some of the many life lessons we learned from Steven Hawking's rich years on our planet.

Don't Stop Believing

When Oxford educated Hawking was first diagnosed with ALS (Lou Gehrig disease) in 1963, he fell into a deep and totally understandable depression. But he found new life in his relationship with fellow student Jane Wide, and despite his dire prognosis, the couple got married and had three children together. Two years after his ALS diagnosis, Hawking received his doctorate from the University of Cambridge. He was elected to Britain's prestigious Royal Society at the age of 32, despite the fact that he was too frail to turn the pages of a book without help. After he was no longer able to talk, Hawking gave his speeches through a vocal synthesizer.

Don't Fence Yourself In

Hawking was almost completely paralyzed, but he visited every continent including Antarctica. He celebrated his 60th birthday in a hot air balloon, gave zero gravity flight a try five years later and hoped to travel to space via Richard Branson's Space Ship Two. He even dared to play poker with Albert Einstein and Isaac Newton on an episode of "Star Track: The Next Generation." In each case, there was probably a chorus of level -headed voices telling him that he had no business doing what he was thinking of doing, but he did it anyway. "I want to show that people need not be limited by physical handicaps as long as they are not disabled in sprit" Hawking said. Lesson learned professor H!

Make them Laugh

What can you say about a theoretical physicist who joked about his IQ on The Simpson's? One of four times he showed on the show, made a cameo appearance in the episode of the "The Big Bang Theory" that was named after him. You can say that Hawking was smart enough to know that laughter is the life preserving gift that never stops giving. Over the years he reeled off quips about sex, morality and Homer Simpson Doughnut Shaped Universe Theory. But when he talked about the importance of humor, he wasn't kidding. "Life would be tragic" he said "if it wasn't funny." Life is an Imperfect Miracle, Cherish It.

Even as a kid Hawking was curious as to how things worked and why. He studied physics and astronomy because he was addicted to wonder, and he stayed that way. Hawking carried that sense of possibility as he made his journey into the center of our universe. And of all the nuggets of wisdom he scattered along the way this one is as a good parting gift as any.

"Remember to look up at the stars and not down at your feet. Try to make sense of what you see and wonder what makes the universe exist. Be curious. And however difficult life may seem, there is always something you can do and succeed at." Stephen Hawking

His ashes were interned between two other great men in London at Westminster Abby, Isaac Newton, farther of gravity and Charles Darwin, father of evolution.

The words man speaks today live on in his thoughts or his memories of others, and the shot fired, the blow strut, the thing done is like a stone tossed in a pool, and the ripples keep widening out until they touch the lives far from ours." Louis L' Amour

The Children

When all the world is young lad
And the trees are green
And every goose a swan lad
And every lass a queen
Then hay for the boat and horse lad
And roam the world away
Young blood must have a course lad
And every dog his day. Charles Kinsley 1819-1875

It is we, the older generation that has done a terrible wrong to those who will inherit the earth after we are gone by leaving them a world that is becoming increasing un-survivable. We knew, or should have known by now that our current and past exploitation of the natural world threatens their very existence. They will be billed for our mistakes and the destruction we have inflected on the earth and then have to endure the physical suffering and heartaches we have left them. The cost of their existence will be impossible for them but they will be compelled to pay it and forced into making terrible decisions about how and where they are going to live their lives.

As we use up the worlds resources there will come a time, not far in the future, when they are gone or diminish and available to only to a privileged few. It is a tragedy that our kids are born into a world where they depend on us their elders to protect them and secure for them a secure future only to have failed them. We are responsible for the climate catastrophe we have left them.

A recent international study by the University of Bath in the United Kingdom found that young people around the world are experiencing "high levels of psychological distress" from climate change and governmental inaction to the growing crisis. Forty five percent of the young people surveyed said anxiety and stress is affecting their daily lives. The study included 10,000 young adults and teenagers ages 16-25 in ten different countries.

Three fourths of those surveyed believe that "The future is frighting" and 65 percent said that their governments were not doing enough to combat the ongoing effects of climate change. "This study points a horrific picture of widespread climate anxiety in our children and young people," said Caroline Hickman, co-author of the study. "It suggests for the first time that the distress in our youth is linked to their government's inaction. Our children's anxiety is a completely rational reaction, given the inadequate response to climate change they are seeing from governments." What more do government's need to hear in order to charge? Greta Thunberg, the young teenage climate activist was quoted as saying, "Young people all over the world are well aware that people in power are failing us."

Kids Suing the State of Montana over Climate Change, IR 7/18/2022

A young lady, Gibson-Snyder and fifteen other Montanan young adults and children are suing the state of Montana. Their lawsuit asserts that the state, by fostering fossil fuel as its primary energy resource is contributing to a changing climate and violating the children's right to a clean and healthy environment guaranteed in the State's Constitution. It claims rightly that the state's reliance on fossil fuels, its energy policy, and its continuing development of fossil fuel extraction has led to the increasing effects of climate change. The state's Attorney General filed a motion to have the children's case dismissed. Four days later the court denied that request by the state. The trial will proceed and is scheduled sometime in 2023.

Helena, Montana: The summer of 2022 was the all-time warmest ever recorded. This past July was Helena's 8th hottest month while last July was the second hottest on record. Helena's five warmest Julys have all occurred since 2003, with 2007 being the warmest. It appears the kids have a good case.

Study: Climate Change disasters occurring more frequency for today's children.

This is the first study of its kind to cover the impact that climate change will have on today's children. If global temperatures continue to rise at their current pace, today's kids will experience in their lifetimes three times the number of disasters that their grandparents experienced. The study published in the Journal Science, concluded that the average 6-year-old can expect twice as many wildfires, 1.7 times as many major tropical storms, 3.4 times more river floods, 2.5 more crop failures and 2.3 times as many droughts then someone born in 1960. Lead author of the study Win Thiery calls it the "intergenerational inequality" of climate change.

Thiery and his 36 colleagues compared the risk faced by previous generations to the number of extreme climate events today's children will witness in their lifetimes. The study shows that today's children will be exposed to an average of five times more disasters then if they had lived 150 years ago. "Young people who will be affected by the ongoing climate crisis are not in position to make decisions" he said. "While the people who can make necessary change happen will not face the consequences of their lack of inaction."

A vast toll of human suffering is hanging over the heads of the world's children, the extent of their pain will be determined on what decisions are made for the control of greenhouse gas emissions this decade. It is the moral duty of today's generation to make the dramatic emission reductions that have been delayed for so long. We will determine today whether our kids grow up in a world with more extreme heat waves, crop failures and chronic food and water shortages.

It is very possible that the numbers given in the report are an understatement. Unfortunately, due to technical difficulties the scientists were not able to assess the increased risk of some hazards such as coastal flooding due to sea level rise. The study did not take into account the increased severity of many events; it only considered frequency. However, this study does make it clear that climate change has arrived. It's everywhere. People today who never considered themselves at risk are waking up to intensified flooding, severe heat and the increasing intensity of numerous wildfires.

Hopefully this study will have a positive effect in that it will bolster the legal efforts of other young people to force climate action on behalf of the children. Last year a Federal appeals court denied a case brought by 21 American young people from a different state that did not have Montana's protection in their constitution. Their case argued that the government's failure to act on climate change was a violation of their rights Similar cases have been filed in other countries.

From CNN News 11/11/2022: Young People Call for Fossil Fuel Non-Proliferation Treaty.
At the Glasgow Scotland World Climate Summit in November of 2021, a group of young people delivered a sharp rebuke to the world's leaders and delegates demanding that a fossil fuel treaty be adopted and calling out world leaders for their support of the coal and gas industries. Several major fossil fuel producers, including

Saudi Arabia resisted that effort. Australia, the largest coal exporter of all the developed nations also fought to opposed the young people and effectively killed their proposal.

"The problem in society is not kids not knowing about science. The problem is adults not knowing science. They outnumber kids 5 to 1. They wield power, they write legislation. When you have scientifically illiterate adults you have undermined what makes a nation wealthy and strong." Neil deGrasse Tyson

"I am angry because this conference is not managing climate change as a crisis. We cannot even include discussions about fossil fuel in the final document. It seems that polluters are more welcome than the people who will be the most affected by these decisions: youth." Said Metiz Jonelle Tan, a young Filipino climate activist. In an open letter addressed to the world's leaders he said fossil fuels were "our generations weapons of mass destruction." They asked for the end of the expansion of any new oil, gas and coal production and a phase out of all existing production. Did they get it? No.

The young leaders shared their anger and sadness about how the summit had progressed. "We already have seen how the biggest delegation at the conference is the fossil fuel lobbyists," Tan said. More than 100 fossil fuel companies sent over 500 lobbyists to the conference. They were by far the largest delegation in attendance.

Dedicated to the young and the significance of change in your world; why should you worry or care?

The oceans are not rising fast.
The severe storms are not affecting me.
Oh, so in 2100 the world will change due to climate change. So, what; that's a long time away.

You are 20, soon faster than know you will be 70. What will your world be like then? Will you really try to understand what's happening now, or will you live one party after another until the party is over and the reality of what is or has been occurring in your world is devastatingly clear.
You live today in a wonderful land of material toys; TV, I-Pads, Twitter and other trinkets. For how long? Realize they are just toys; new ones will come along and today's amusements will fade. Your world is warming at a rate not experienced in the last 10,000 years. At least try to understand what is happening and possibly dedicate yourself to do something about it.

To waste, to destroy our natural resources, to skim and exhaust the land instead of using it so to increase its usefulness, will result in undermining in the days of our children, the very prosperity which we ought by right, hand down to them. Theodore Roosevelt

The Oceans and Fresh Waters

Unsustainable exploration, pollution and the climate changes threaten marine species and ecosystems in all the world's oceans. Less than 3% of the global oceans have escaped human pressure. This will need to change otherwise there will be profound impacts to the ocean's ecosystems. Severe weather systems will increase and the more than three billion people whose livelihoods depend on the oceans and its coastal biodiversity will be seriously impacted. These costal marine and polar ecosystems must be restored and maintained and the use of their resources made sustainable.

More than 90% of the heat trapped by carbon emissions in the atmosphere is absorbed by the oceans making their warming an undeniable signal of an accelerating crisis. Research has found that the five warmest ocean episodes have occurred since 2015 and the rate of heating since 1986 was eight times higher than that from 1960 to 1985. Warmer water is less able to dissolve carbon dioxide thereby reducing the amount of carbon the oceans can absorb so more is retained in the atmosphere increasing heating.

Over the past century sea surface temperatures have risen by an average of about 0.13 degrees Celsius (0.23F) per decade as the ocean has absorb vast amounts of greenhouse gas emissions from burning fossil fuels. If this trend continues for the remaining 78 years of this century, without any additional increase in emissions, it would amount to a rise of about 1.8 degrees Fahrenheit. That doesn't seem like much but just small increases in water temperatures can have profound effects on marine species. These temperature increases and the changing ocean chemistry affect all sea life including essential parts of the food chain, they also cause fish populations and other marine species to migrate to cooler waters.

Warmer oceans supercharge the weather, impacting the biological systems of the planet as well as the human society. The study also reported that the sinking of surface ocean water and upwelling of the deeper cold water is reduced as the seas heat up. This means as the surface layers heat up fewer nutrients for marine life are circulated to the surface to sustain the ecosystem.

Marine heat waves which intensify with climate change, are rapidly emerging as forceful agents of disturbance with the capacity to restructure entire ecosystems. With confidence growing among scientist in their ability to project extreme warming events due to the changing climate, marine protection must consider the ocean's heat waves and these intense climate events if we are to maintain and conserve the integrity of the highly valuable marine ecosystems over the coming decades.

Due to increased greenhouse gas emissions, the extended periods of extreme warming in the oceans have increased in frequency and intensity by 50 % in just the past ten years. These marine heat waves destroy the biodiversity and ecosystems of the ocean, increase the frequency of severe weather events and impact the fishing and tourism industries. They have been recorded in the surface and deep waters across all areas of the world's oceans and in all types of marine ecosystems. Ocean heat waves are defined as when areas of the ocean reach extreme temperatures that continue for five days or longer. As these events become more frequent and severe, they risk pushing the oceans beyond their ability to recover. The damage caused by these "hot spots" is harmful for humanity and all species that rely on the oceans for their ability to produce oxygen, food and the removal of climate-warming carbon dioxide from the atmosphere. This will have lasting consequences for the oceans biodiversity and the many millions of people whose livelihoods depend on these systems.

It has been recorded that heat waves have killed or reduced the productivity of economically important species such as lobsters and snow crabs in the Atlantic and scallops off the coast of Western Australia. Fish, whales, birds, kelp, sea grass and plankton along with many other species can be affected by the increased temperatures. They have been associated with the mass mortality of marine invertebrates; they force species to change their behavior by migrating to cooler waters, creating harmful conditions.

Newsweek, Toxic Algae: The California Water Quality Monitoring Council has detected a severe algae red tide in San Francisco Bay as well as "dangerous levels" of algal toxins in Clear Lake, Lake Crowley and Bridgeport reservoir. Algal blooms are caused by rapid growth of algae due to increased nutrient content of the water and by increased water temperatures. According to the Center for Disease Control and Prevention these toxins may be absorbed by the skin, inhaled, swallowed, or consumed in contaminated foods. If exposed, humans may experience flu-like symptoms, skin irritation, paralysis and abnormal breathing. In pets', exposure may lead to seizure and death.

Algal blooms can have catastrophic effects on aquatic ecosystems. Their rapid massive growth uses up all the oxygen in the water, starving marine life of oxygen as well as producing harmful gasses such as CO_2 and toxins that can kill other organisms. If the algae become too dense, they block out the sunlight to the water below, causing marine plants to die. The red tide algal bloom occurring in San Francisco Bay, the largest of its kind in over ten years, is killing thousands of fish. According to the California Department of Fish and Wildlife as many as 10,000 fish died in late August alone. The bloom on the Bay is thought to have been caused by increasingly warm waters and excessive sewage providing the algae with extra nutrients. Algae blooms are expected to get worse with the warming caused by climate change, both in fresh and salt water environments.

Snow Algae. Red-pigmented green algae are found in high alpine and polar regions around the globe. Scientists have established that snow algae could play a major role in the melting of snow fields and glaciers in the future. Because of its brightness snow reflects the heat of sunlight back into the atmosphere reducing melting. When the dark algae get established in a region it darkens the snow and absorbs the suns heat, increasing the rate of melt. This can start the process of a feedback loop where increase heat produces more algae that increases heating and more rapid melt. This feeds the snow algae with fresh water and increased nutrients and more sunlight. Thus, the algae alter its own habitat and the adjoining areas.

The rate of snowmelt is critical for draught areas of the world. If snow algae hasten snowmelt or melts all of the snow quickly, streams end up warmer than usual, with less water during the warmer summer months. In 2021 an article in the journal of Nature Communications found that algae blooms were responsible for as much as 13% of the surface melting on Greenland's ice sheet, another study in Alaska reported that snow algae accounted for as much as 17% of the total melting rate on one large icefield.

Approximately 71 percent of the earth's surface is covered by water. The oceans contain about 96.5 of all the earths water and that water has had an increase in temperature approximately 1.5 degrees Celsius in the last century. For the past ten years annual warming temperatures increases have been the highest ever recorded. Higher ocean temperatures can escalate the number and intensity of tropical storms and hurricanes. They disrupt the water cycle and intensify the possibility of increased flooding, droughts and wildfires and algae growth. They also heighten the stress on the oceans ecosystems by increasing stratification, acidification, and deoxygenation of its waters.

The rate of sea level rise has doubled since 1993 according to the World Meteorological Organization. "The past two and a half years alone account for 10% of the overall rise in sea levels since satellite measurements started 30 years ago. NASA and France's space agency Central Natural started together flying satellite altimeters in the early 1990s, beginning continuous space-based coverage. That effort continues with the 2020 launch of the joint U.S.-European Sentinel-6 satellite with its altimeter which will provide scientist with uninterrupted satellite records of sea levels for over three decades.

The United States is expected to experience as much sea-level rise by 2050 as in the previous 100 years. Sea level rise is accelerating along the U.S. coast and is expected gain an additional increase of up to 6 inches by 2050, according to a new National Oceanic and Atmospheric Administration (N0AA) study. It will continue to rise as the worlds melting glaciers and Greenland and Anthracic ice sheets catch up with todays heated atmosphere and future warming. That will double the amount of sea level rise that has occurred over the last century. Such an increase would threaten cities such as Miami, Boston, and New York, where flooding is already occurring during the highest tides and storm surges. While the amount of rise will vary from location to location, the new data is "code red" for the deepening climate emergency, said Gina McCarthy, NOAA's National Climate Advisor. Sea level rise is a clear and present risk to the United States today and for the coming decades and centuries the report stated.

The primary cause of sea level rise will be the warming climate that is melting vast amounts of ice from the world's glaciers and Greenland and Antarctic ice sheets that contain over 60% of the earths fresh water. These ice sheets are massive, some more than a mile thick. As the atmosphere warms the melting will be slow, a progression that has already started.

A simple example of what is occurring can be described as follows: Set in a room (the planet) where the temperature is 70 degrees (the atmosphere), now place a 50-pound chunk of ice on the floor (the ice sheets). We know in time that the ice will melt and by varying the temperature we can increase or decrease its rate of melt. It will only stop melting when the temperature is maintained at the freezing point of water, 32F degrees or less. So as long as there is a continuing warming trend, it doesn't have to be constant, melting will continue. Obviously the larger the ice chunk, the longer will be the melt rate; it will take considerable time to melt the ice sheets, hundreds of years, but melt they will, and the oceans will continue to rise. We must remember that the melting of these ice sheets has occurred in the past during earth's long history when the warming was similar to what is occurring today. Even a thousand years of melting means nothing to Mother Earth; but what does it mean to mankind?

A new study in the journal Science examined all of the globe's 215,000 glaciers not including the ice sheets of Greenland and Antarctica. They used computer simulations to calculate how many glaciers would disappear, how many trillions of tons of ice would melt and how much sea level rise would occur. According to the study the world is on track for a 4.9-degree Fahrenheit temperature rise by the year 2100 which will result in losing 32% of the worlds glacier mass and 68% of its glaciers. Using the above project temperature rise this would increase sea levels by 4.5 inches not including the melting ice sheets and effects of expansion due warmer water. Project glacier ice loss by 2100 could range from 38.7 trillion metric tons to 64.4 metric tons depending on how much the world warms and how much coal, oil and gas is burned.

The 4.5 inches of sea level rise from just the glaciers would mean that more than ten1million people around the world and more than 100,00 people in the U.S. would be living below the high tide lines. Scientists say that the major future sea level rise will be determined primarily by the continued melting of the

Greenland and Antarctic ice sheets. The loss of the glaciers means more than just rising seas. It will result in shrinking water supplies for a large part of the worlds population and more severe flooding as they melt. "No matter what we do we are going to lose a lot of glaciers," said lead author David Rounce a glaciologist and engineering professor at Carnegie Mellon University.

What's at stake are thousands of homes, as well as airports, military bases, seaports, power plants, oil refineries, bridges and highways and city infrastructure that eventually will have to be relocated and rebuilt. At what cost? The U.S. will get slightly more sea-level rise than the global average The greatest rise will be on the East Coast and Gulf. It will include areas that haven't flooded in the past, many of our major metropolitan areas on the East Coast are going to be at increased risk

By 2050 disruptive and damaging sea level rise and increased storm and tidal action will occur more than ten times as often as they do today. Said Nicole Le Boeuf, NOAA's National Ocean Service Director. Only by reducing emissions can these events be limited. Nothing else is going to dampen sea level rise and the world isn't going to suddenly cool down, there are no natural cycles that will save us.

The following numbers indicate five possible sea level rise consequences for the globe and the U.S., measured in meters.

Year	Global			United States		
	2050	2100	2150	2050	2100	2150
Low	0.15	0. 30	0.40	0.31	0.60	0.80
Intermediate Low	0.20	0.50	0.80	0.36	0 70	1.2
Intermediate	0.28	1.0	1.9	0.40	1.2	2.2
Intermediate High	0,37	1.5	2.7	0.46	1.7	2.8
High	0.43	2.0	3.7	0.52	2.2	3.9

Carefully observing the above projections of possible sea level rise even at the "Intermediate Level," at the end of the century in the U.S. the project rise is 1.2 meters or approximately 3.6 feet. At the worst "High" it would be almost 7 feet. Nowhere does data show a stabilization, the rise continues to increase. So, what happens after 2150 as the planet continues to warm? Only a hundred and twenty-eight years in the future; can we think and plan that far ahead? We better.

The sea ice surrounding Antarctica reached its lowest extent on record for July according to satellite data from Europe's Copernicus Climate Change Service. A record low coverage was also reached in June, following several months of below -average coverage. The service said Antarctic Sea ice covered just 15.3 million square miles in July, or 7% below the 1991-2020 average for the month.

The Antarctic ice sheet is the largest block of ice on Earth. It covers more than 5.4 million square miles and contains 7.2 million cubic miles of fresh water. This ice sheet averages about 1.2

miles thick and if completely melted would rise sea levels more than 200 feet. It has happened before in Earth's long history with its melt water added to the ocean that created vast areas of large flooded inland seas. During a warm period, some 125,000 years ago massive amounts of Antarctic's ice melted rising sea levels by some six to nine meters (18 to 27 feet). This melting happened over a time period of several hundred years.

The Greenland ice sheet is much smaller than Antarctic's, it covers about 656,000 square miles. It is the second largest chunk of ice on the planet. According to NASA, 5 trillion tons of ice has melted off this ice sheet during the past 15 years. Both ice sheets have been losing vast amounts of ice at an increasing rate since the 1990s. There is a degree of uncertainty as to how fast these sheets will decline in the future. It depends on how effective the efforts are to reduce the amount of additional greenhouse gases emitted into the atmosphere. Studies have shown that Greenland's ice sheet has become very sensitive to climate change. A large amount of future melting of both ice sheets unfortunately, has already been assured due to the enormous amount of the existing gases that will stay in the atmosphere and continue to heat the planet for hundreds of years. These ice sheets have started melting now, due to today's temperatures and they will continue to melt, the only question is the rate of melt. If we were able to eliminate all future emissions the melting would still continue. It may be slow and uneven, but uninterrupted.

Sea Level Change: More than 220 individual glaciers flow from Greenland's ice sheet into the ocean. In December 2021 NASA completed a six-year study of these melting glaciers. Its purpose was to determine if the ocean water was melting the glaciers as much as the warming air is melting their upper surface. The melting of these glaciers currently contributes more to the global sea rise than any other source. The findings of the study revolutionized scientists' understanding of the pace of sea level rise in the coming decades. They found that vast areas of the shoreline glaciers are being undercut by the oceans warming waters creating conditions that will accelerate present and future break up. This clarified the likely progress of future ice loss in a place where glaciers are melting six times faster than they were only 25 years ago. If all of Greenland's ice sheet were to melt global sea levels would increase by about 24 feet. We can only guess how rapidly that might occur or how much ice loss will happen before conditions stabilize if they ever do; surely several hundred years. In the mean time they will continue to slowly melt, and sea levels will continue to increase.

There will be secondary effects due to the amounts of fresh water pouring into the ocean. These vast amounts of fresh water may lead to the reduction of krill in the waters around Antarctica; krill are the foundation of life and food in that area. So as melting continues it creates changing events to the ecosystem's even before substantial melting occurs. Without krill, as without plankton in the warmer waters, entire ecosystems can collapse.

Fresh Waters

We humans and all of our living land companions; our flora and fauna depend on the availability and access to fresh clean water. No single element on earth is more important or valuable. We could have pockets full of gold but what would it really be worth if we really needed a drink of water? Yet, for a long-time humanity has damaged this precious resource and our freshwater ecosystems are disappearing. Fresh water is the lifeblood of our planet. It is essential for our very existence, our health, our food and much more. But too often we take it for granted. We waste it, pollute it and ignore its necessity in our lives.

Only 1% of the Earth's surface area is covered by fresh water and only 0.3% of that water is available as surface water in our streams, lakes and rivers. Most of the fresh water is captured underground or in glaciers and polar ice caps. Greenland and Antarctica ice caps hold the most at 69 %, groundwater accounts for 30% and again surface water amounts to only 0.3%.

Many countries are today facing serious fresh water shortages in the face of a growing world population. Climate change, with the increasing heat, is impacting the availability and necessity of water to millions of people around the world. By 2030 it is projected that there will be a 50% increase in the demand for water. The United Nations projects that over 40% of the world's population will live in severely stressed water locations by 2050. By this time there will be 9 billon people on earth who will require 60% more food, 55% more water and 80% more energy. Where will all of these necessary resources come from?

Residents of Monterrey, the capital of the prosperous state of Nuevo Leon in northern Mexico recently had running water for only a few hours a day. In some of the poorer districts it has been over fifty days since the residents saw any water in their faucets. Several reservoirs serve the city. The water in one had dwindled to less than one percent of capacity by late June while another was at 7% and a third at 44%. There have been fifteen months of scant rainfall. Farmers and the booming industrial sector that is dominated by the production of soft drinks, beer, steel, and cement have been subject to few water restrictions despite the drought. The citizens of the city have seriously resented this unequal treatment.

The consequences of badly managed water resources are becoming severe. A recent report by sixteen conservation organizations warned that as many as one third of all freshwater fish face extinction by the end of the century. Habitat destruction, dams, over use, agriculture and industrial pollution are causing fresh water biodiversity to decline at twice the rate of land or marine species.

Groundwater: Scientist have discovered that the dissolved organic matter in groundwater, when it is brought to the surface, can easily be turned into carbon dioxide. Unfortunately, that means that groundwater is likely to be another source of planet heating global warming. As the world warms and available surface water decreases the extraction of ground water by deep drilling will increase. Deep groundwater contains more carbon containing dissolved organic matter.

The major source of carbon producing groundwater is not that pumped to the surface. Every day subterranean groundwater seeps out of the world's coastlines into the oceans at a rate of about 10 time the amount being pumped. The carbon, oxygen and hydrogen molecules containing the dissolved organic matter are then converted by sunlight into carbon dioxide (CO_2) that eventually flows into the atmosphere. At present this carbon contributed by groundwater as a climate change source is not included in the global carbon release estimates. It should be factored into the way we deal with climate change. It is estimated that 12.8 million tons of dissolved matter is brought to the surface or flow into the ocean every year.

Plastics, Chemicals and Pollution

Plastic pollution now affects almost every species living in the ocean and in lakes, and many on the land including us. At the same time, the high sea surface temperatures once considered extreme have now become the new normal. These are the conclusions of the two recent studies published ahead of the One Ocean Summit, a conference organized by French President Emmanuel Macron. The summit was presented to establish guide lines to protect marine life from overfishing, climate change and pollution. The research papers tell a story of an ecosystem vital to human survival that is increasing under threat.

The study on plastics, organized by the World Wildlife Fund and conducted by scientist at Germanys Alfred Wegener Institute found that regions in the Mediterranean and East China Sea had critical plastic pollution. It documented the negative impacts of plastic on 88% of 297 species studied. Plastic pollution is now found almost everywhere in the world's oceans and is almost impossible to remove. Even if the additional plastic pollution could be stopped today the amount of microplastics would double as the larger pieces break down. Microplastics are defined as the fragments of any type of plastic less than five millimeters (0.20 inches) in length. Once plastic has started to decompose into microplastic or nanoparticles it can no longer be removed, it is too small.

Ocean currents are sweeping up huge quantities of plastic waste into what is called ocean garbage patches and one of the largest is the Great Pacific Garbage Patch that occurs northeast of the Hawaiian Islands. It's estimated to cover an area twice the size of Texas, containing trillions of pieces of plastic debris including microplastics that are harmful to sea life. To date, no country has made a firm commitment to clean up this increasing mess. We spend billions of dollars playing in space and on the military establishment but very little preserving our home; earth.

Microplastics may significantly intensify global warming according to a new study. Scientist from different German research institutions, including GEMAR Helmholtz Center for Ocean Research in Kiev, have found that up to 27.5 million tons of microplastics are being transported thousands of miles all over the world by oceans and land by wind, rain, snow, sea spray and fog each year. The new study shows that wind can transport the particles great distances. They can travel from their point of origin to the most of the corners of the planet within days.

The team of 33 international researchers warns this could affect the surface climate and health of local ecosystems; as an example, when dark microplastic cover snow and ice it reduces their ability to reflect the suns energy back into space promoting faster heating and melting as a consequence.

The investigation also shows that microplastic particles can serve as condensation cores for water vapor. By doing this, they affect the formation of clouds and in the long term the climate.

The ocean has become the trash can for plastic waste on the planet. Today, between 19 and 23 million tons of plastic litter ends up in the world's oceans every year. Only 9% of all the plastic made has ever been recycled, about 12% has been burned and the rest ends up in landfills or is left lying around that eventually gets into waterways that carry it to the sea, inland lakes and rivers and streams.

"Plastic ends up in our food more often than many people would suspect, in fact I'd say that almost everything we eat today has traces of plastic," said Jane Munake of the Food Packaging Forum, a nonprofit foundation that studies chemicals in all food packaging materials and their impact on health. "This is either tiny bits of microplastic or plastic chemicals. Both can originate from plastic food packaging, or food processing equipment, or sadly from environmental pollution. Plastics are everywhere and they are persistent in getting into our food where we might ingest them," she said.

The problem is that we do not have good options for disposing of plastic waste. Trillions of tons of plastic are produced each year and they exist long after we have finished using them for a few seconds or minutes. As you read this, look around and observe how many things you observe contain some form of plastic. Some may pose a serious threat to humanity today and in the future as chemicals in plastic have been linked to dropping male fertility rates, cancer and other serious health ailments. They are also harming the environment with sea life ingesting microplastic, wildlife getting entangled in plastic and whales dying; found with their bellies full of plastic junk.

From the moment we are born, if not before, we are exposed to the effects of plastic and we don't really know what it is doing to our health and wellbeing. A recent investigation by a team of researchers in China has uncovered signs that elevated levels of microplastics could be inflaming our digestive system. Feces collected from 52 individuals diagnosed with inflammatory bowel disease (IBD) were found to contain 1.5 times the number of plastic particles then samples taken from those without any chronic illnesses. It was found that the greater the amount of plastic the more severe were the IBD systems. It should be understood that this was just a small observation study and did not establish conclusively the direct link of the plastics with IBD. Further studies will be required but this is another level of concern about the increasing amount of plastic in our environment.

The United States alone in 2016, generated 46.3 million tons of plastic. That amounts to 287 pounds per person in just one year.

Surely that amount has increased since then. It is estimated that plastic use will more than double worldwide by the year 2050. Is it possible that we can drastically reduce its use before then, and if not where will all that used plastic go? And how much oil and energy will be necessary to produce it? A report by the environmental organization, Oceana found that the plastic waste from Amazon's packaging increased 18% last year. According to the report Amazon's plastic waste increased from 599 million pounds in 2020 to 709 million pounds in 2021. This is an example of just one company, there are thousands more like it around the world.

Plastic is toxic at every stage of its life cycle, from the volatile chemical compounds used in its creation to its disposal. Many of the synthetic substances are, in the words of one study, "difficult or impossible for nature to assimilate" and they degrade the environment. At least 2400 of the chemicals commonly used in the manufacture of plastics, according to a study in the journal Environmental Science and Technology, are considered "substances of potential concern" by the European Union's chemical governing body, because of their toxicity, persistence to stay in the environment or their capacity for accumulating in human or animal bodies. At least 35 have earned the EU's highest levels of concern due to their status as being "very persistent in the atmosphere;" many hundreds of chemical substances commonly used in plastic production have not been studied.

Plastic producers are also potent greenhouse gas emitters; The more than 30 ethane producers in the U.S. together released 70 million tons of greenhouse gasses in 2020, according to the research firm Beyond Plastics. That release amount is equal to the annual emissions of at least 35 average- sized coal-fired power plants. The International Energy Agency predicts that by 2030 petrochemicals will account for more than a third of the growth in the world's oil demand. Petrochemicals are produced from oil and are used in the manufacture of many products, a principle one is plastics. So, it isn't just coal that we should be concerned about.

A new petrochemical chemical plant recently completed in Texas by Saudi Basic Industries (SABIC) in a joint venture with America's biggest energy company, Exxon Mobil. SABIC is a $40 billion company that manufactures chemicals, fertilizer and plastics and is owned by Aramco, the world's largest oil company that in turn is owned by the House of Saud, Saudi Arabia's royal family.

Their new factory in Texas, through a process that heats hydrocarbons, creates ethylene. Ethylene is then converted into polyethylene the basic ingredient of tiny plastic pellets, called nurdles. These are the pre-production plastic pellets that are heated and molded into a variety of shapes to create products such as plastic bags and beverage containers. They become the single use non-biodegradable products that end up landfills, the oceans and fresh water environments. SABIC owns seven other plastic manufacturing plants in five U.S. states. Many of these plants reported leaks of chemicals associated with cancer, fetal mutations and respiratory ailments according to the EPA Toxics Release Inventory (TRI). Between 2016 and 2020, SABIC facilities in the U.S. producing plastic resins released an average of 120,500 pounds of styrene each of those five years. Styrene is classified as a possible human carcinogen by the World Health Organization International Agency for Research on Cancer (IRAC) and can cause memory loss, confusion and slowed reflexes. Altogether, between 2016 and 2020 the company's facilities in the U.S. were among the top emitters of at least seven toxic substances. Those substances included BPA and butadiene, by-products of plastic resin production and classified as known human carcinogens.

Over the last decade, SABIC has paid over one million dollars in fines for violating EPA and the Occupational Safety and Health Administration regulations. At the Texas plant since it open in December, it has had eight emission events of unauthorized releases of air contaminants. On June 16th, 478 pounds of nitrogen oxides, a contributor to respiratory disease was released. In May it had released 572 pounds of benzene, a known carcinogen. A study by the Environmental Integrity Project revealed that even in cases of clear violations the states enforcement agencies rarely enforce emission limits. Less than 3% of 500 million pounds of illegal air pollutants resulted in penalties. Jane Patton, an organizer with the Center for International Environmental Law, an organization that uses litigation to force compliance with environmental laws, said "the regulatory bodies in Texas and Louisiana are little more than rubber stamps for the oil and gas industries."

These hundred-billion-dollar companies are so powerful that a million-dollar fine is a pittance. It is much cheaper for them to pay the fines, if any, and continue breaking the law by releasing huge amounts of pollutants into the atmosphere, endangering people's health and lives, then to comply with the regulations.

As an example of the inclusion of plastic into our environment: ***Sampling Finds Plastic Pollution in Montana's Fresh Water.*** A study was made by the state's environmental laboratory that showed that plastic pollution was found in more than half of the 50 water samples taken from lakes and rivers around the state. The goal of the study was to determine the presence and types of microplastics near fishing access sites. Ten quarts of water were taken at each site: all jars were labeled and numbered, with site description and date. The identified microplastics were categorized into four types for laboratory testing.

1. Fibers from synthetic fibers and filaments such as fishing lines and bailing twine.

2. Fragments from rigid plastic, including polystyrene and clear plastic containers.

3. Films from plastic bags and food wrappers.

4. Micro beads from older personal care products.

Of the 50 sites tested, 33 (66%) contained one or more types of microplastic. That is happening today in Montana. Is the same thing happening in the other states and what will it be the results in say fifty or a hundred years?

Another example: A new study by the University of Montana Flathead Lake Biological Station found that wind, streams and land surface runoff carry microplastics into Flathead Lake. They are accumulating in the water creating a toxic threat to humans and ecosystems. This is occurring not only in the oceans but in all the worlds waterways creating a global issue. Microplastics first found in seafood now are found in most places' where humans live, reported station director Jim Elser. Flathead is not the only location they are found; many rural lakes contain them. The problem is caused by the sheer number of plastics we Americans and the **rest of the world use. They will continue to build up in small amounts and when they degrade** and can't be seen they will still exist.

Blood Plastics: Dutch researchers say that for the first time, microplastics have been found in human blood samples. Writing in the journal Environment International they document how a small study detected traces of the plastic pollutants in the blood sample of 17 out of 22 volunteers. The study suggests the PET plastics and polystyrene particles were likely inhaled or ingested before ending up in the bloodstream. These substances are found in plastic bottles, polyester fibers, and other products. The researchers also emphasized that more studies are needed to determine if the man-made substances pose a public health risk.

The incredible diversity in the deepest levels of the ocean is being threatened with extinction by microplastics. These findings were recently discovered after scientist studied the impacts of pollution in the West Pacific Kuril-Kamchatka Trench in the Pacific Ocean. They discovered with core samples taken from the ocean sediments at a depth of over 6 miles that the level of microplastics were far higher than expected.

Researchers, Serena Abel and Angelika Brandt from the Alford Wagner Institute Center for Polar Marine Research, Goethe University, Frankfurt, Germany took 13 samples of deep-sea sediment from different locations. They discovered that the samples contained between 215 and 1,596 microparticles of plastic per kilogram of sediment. Not a single one of the samples were free of microplastic. Their research found a total of 14 different types of plastic. Using a special spectrometer, they were able to detect even the smallest of the microplastic particles.

Microplastics were recently found in what was thought to be a pristine area in the mountains of the French Pyrenees at an elevation of 9,440 feet. The study of air movements indicated the particles originated in Africa, North America or the Atlantic

Ocean which indicates that they can be distributed over very long distances. They have been located in the deepest oceans and the highest mountains; it may be that there are no more pristine areas left on the globe.

Toxic World: A new study suggest that chemical pollution has become so pervasive that it has pushed the Earth outside the relatively stable environment of the past 10,000 years. Beside the widespread use of plastics, researchers say they are highly concerned about 350,000 synthetic chemicals, including pesticides, industrial compounds, and antibiotics. "There has been fifty-fold increase in the production of chemicals since 1950 and this is projected to triple again by 2050," said research team member Patricia Gomez of Sweden's Stockholm Resilience Center.

Forever Chemicals: Wildlife agencies in several U.S. states are finding high levels of a toxic chemical in game animals and in some cases fish. They have detected elevated levels of PFAS or polyfluoroalkyl in deer in Michigan and Maine. Sometimes called "forever chemicals" because of their persistence to stay in the environment. PFAS are industrial compounds used in numerous products. They are linked to health problems including cancer and low birth weight.

The discovery of these chemical in wild animals hunted for sport and food represents a new challenge and some states have issued "do not eat" advisories for deer and some fish. Degrading very slowly or not at all the chemicals can remain in the bloodstream for a lifetime. They enter the environment through the production of manufactured goods; cosmetics, medical products, firefighting foam and agriculture. PFAS tainted sewage sledge has long been applied to farm fields as fertilizer for many years. Scientists have found significant levels of "toxic forever chemicals" in rainwater collected around the planet. Even in Antarctica they discovered PFAS at 14 times higher than the level the Environmental Protection Agency considers safe.

More than 20 states have proposed or adopted limits for the chemical in drinking water. In Maine where the chemical was detected in wild animals, a well was tested and found the water contained hundreds of times the federal health advisory level. Wisconsin has tested deer, duck, and geese for PFAS and as a result issued a "do not eat" warning for deer liver in certain locations in the state. It was also recommended that fisherman reduce their consumption of Lake Superior's rainbow smelt to one meal a month.

It has been found that bees, butterflies, and other pollinators exposed to common air pollution are significantly impaired their ability to sniff out plants that depend on them for pollination, according to new field research. British scientist says that the pollution, combined with land use changes, are responsible for up to a 70% drop in the number of pollinating insects. Writing in the Environmental Pollution, the team said they exposed a test field to levels of pollution commonly found near highways and observed up to 90% fewer flower visits by pollinators.

Plastic pollution in the Arctic is as bad as anywhere else on earth. The snow and ice are polluted with microplastic carried north by the wind, currents, and waves. High concentrations have been found on the seafloor and remote beaches. Found in all marine organisms from the smallest plankton to the largest whales, it accumulates in their bodies and tissue through their food chain. It is quite possible that the long-time ingestion of microplastics in the seas fish and mammals could lead to their reduced growth and reproduction along with producing increased stress and inflammations. These are the conditions occurring today. What of the future, how much additional plastic can the oceans and their inhabitants consume before they up-chuck and become really sick?

Air Pollution: Levels of particulate matter in California, Oregon, Nevada, Idaho, and Washington have become more than twice the recommended safe limits mostly caused by wildfire smoke. Particulate matter is a mixture of microscopic solids and liquid droplets that are 2.5 micrometers in diameter, or roughly one - thirtieth the width of a human hair. Commonly found in wildfire smoke and vehicle exhaust, especially diesel exhaust, they have been connected to respiratory-related hospital admissions and increased mortality. When you inhale these particles, they are able to penetrate deep into the respiratory system and once these particles enter the lungs, they can cross over into the bloodstream and enter other organs.

The U.S. sewer systems were not built for the effects of today's climate change. As more intense storms occur with increased rainfall, many communities with antiquated sewer systems that have combined rainwater, snowmelt and toilet waste piping systems are having serious problems with sewage backup in their homes. The sewage and surface drainage systems were never designed to carry the discharge of rainfall amounts that are now occurring. When the city's sewer system reaches its capacity if overflows into streets, flooding basements with raw sewage. Throughout the country there are 728 communities that have these combined systems. Most large cities have a dual system with separate drainage for surface water and sewage; they are not mixed.

Because these aging systems have not kept pace with new developments and population growth, the combined systems spilled 850 billion gallons of untreated raw sewage into open waters, rivers, and streams in 2004, the last time estimates were made. Rife with feces, pathogens, debris, and toxic pollutants these flows create a serious risk to human health and the environment. The discharges contaminate drinking water and increase algae blooms in rivers, lakes and salt water environments. It will get worst as climate change creates more heavy rainfall.

As an example, one week in early September 2018, a pair of severe storms swept across the Midwest and Northeast swamping many small communities that had the combined systems. Each city was overwhelmed, discharging a total of 330 million gallons of raw sewage-laden water into nearby rivers. Addressing climate change and its impact on water infrastructure will be an expensive undertaking if we are to avoid continuing damage to the environment and human health.

Colorado, Denver: Researchers are finding increasing amounts of microplastics in the state's snowpack sending even more pollution into the state water supply.

The water quality in Chesapeake Bay has only gotten more damaging according to data from a study that found more dead zones and algae blooms in the bay's tributaries.

Energy

The scientists say that 85% of matter is dark matter and they don't even know what that is. Well, if is the most abundant thing in the universe, it has to be made of stupidity. Dilbert cartoon

Fossil fuels account for more than 75 % of all global energy use and they are also are essential for the making of thousands of other products. Most all plastics are manufactured using fossil fuels. Will the rush for profits and political pressure override the need for immediate and drastic climate action? How will humanity change course in time and will we be able to survive without our pickup trucks, motor homes, busses, trains, aircraft, pleasure crafts, seagoing commercial transport, plastics and thousands of other uses and products that require fossil fuels, that we rely on today? We have been caught in a serious predicament, speared by our own petard; dammed if we continue down the path we are on and dammed if we don't.

Despite the need to make significant reductions to greenhouse gas emissions, many energy producing countries are actually projecting an average increase of fuel production of 2% annually until 2050. We must realize that future large-scale development and use of fossil fuels is no longer suitable for the planet. The government must reduce the development of these fuels to a minimum that is necessary to maintain supplies for critical infrastructure. Off shore leases must be curtailed. There should be no more government handouts to the oil industry. The taxpayers should no longer be subsidizing the industry by $7.6 billion dollars per year of taxpayer's money with tax breaks, direct payments and public finance. To control this, we need a tax on all carbon emissions to force the industry to invest in renewable energy by putting a price on all emissions.

Unfortunately, oil and gas will still be required in large quantities. We cannot lubricate our industrial world with sun and wind energy. But we certainly must control the amount of these non-renewable sources of energy that are being slowly depleted. Diesel will still be necessary, there are trains and ships and diesel trucks to run. Large quantities of kerosene will be required for aviation fuel. Unfortunately, there is no free lunch, the world has reached a real dilemma. How do we continue to control the heating of the planet and maintain a stable economic environment? And another question arises: How much longer will these resources last? We have been burning them up at a prodigious rate over the last one hundred years; will they last another two, three, five hundred years into the future? Very short time spans if we expect to still be around.

There have been expensive attempts to develop "clean coal." There is no such thing we have belatedly learned; billions of dollars have been spent on research to develop clean coal and nothing has come of it. The demolition (blown up) of the country's largest plant devoted to that effort was just recently completed. It was 2.4 billion dollars in debt after years of propaganda as to how successful the effort was.

Germany: After a major gas pipeline from Russia to western Europe was shutdown, Germany prepared to give the green light for 10 coal-fired power plants to restart among concerns Russia may not resume the flow of gas. Coal burning around the world for power production is scheduled to reach a new high this year as the increased demand has intensified due to the war in the Ukraine.

Species Extinction and Forest Reduction

Extinction is the rule, survival is the exception, 99% of all species that ever existed on earth are no more. Carl Sagan

We cannot command nature except by obeying her. Francis Bacon

Men and civilizations are all alike. They are born, they grow to strength, they mature, grow old and die. It has always been that way down through Earth's history.

Are we humans the "frog in the pan?" Do we lack the capacity to feel the water slowly boiling (climate change) and jump or are we destine to boil gradually away in our own caldron? As we rightly increase wind and solar power do we decrease the overall benefits by promoting growth for growths sake? When will we attack all emissions? Until we do it's a zero-sum game and the earth will lose.

From the Journal Science Sept. 20, 2021. The same greenhouse gas that is threating the world today was responsible for the largest mass extinction in the planet's history. The end- Permian mass extinction occurred some 252 million years ago when over 90 % of all marine life and over 70 % of all life on land were destroyed. It was caused by the release of a deadly substance into the atmosphere by volcanos. Volcanic eruptions can have a massive effect on Earth's climate. The deadly material was carbon, the same greenhouse gas that is threatening the world today was the cause of this mass extinction, the largest extermination that has ever occurred.

This has recently been discovered by researchers at Montclair State University in New Jersey. It was found that an immense amount of carbon dioxide was released from volcanic sources within a short time period of 15,000 years. It raised the average temperature from 25 degrees Celsius to 40 degrees. The 15 degrees increase was deadly to most life on Earth at that time. For comparison, in a period of just the last two hundred years, carbon emissions due to human activity have resulted in an average temperature rise of over 1 degree Celsius across the world. Only a one degree increases in 200 years, yet the world today is experiencing the severe climate upheavals that we see every day.

The researchers studied carbon isotopes from samples collected from sediments extracted from the Finnmark Platform in Norway on the eastern part of the Barents Sea Shelf. The sediment samples were well- preserved and allowed the researchers to look at biomarker compounds from life on land and oceans during that period of Earth extinction.

The Earth is heading towards 4 degrees Celsius (4C) warming based on the emissions reduction commitments recently made by the world's government. These commitments are insufficient to slow down the planets warming unless more serious and concrete actions are taken. The current path leads to 3.3C by the end of the century but that does not include the carbon feed-back that have already become active; the permafrost melting, the continuing destruction of the Amazon rain forest and the oceans declining ability to absorb carbon due to increased heating. These conditions would extend the heating to 4C.

We must remember that since the last mass extinction, 65 million years ago, the earth had been reasonably stable. There has been ups and downs, a couple of severe ice ages that covered large segments of the globe, and huge heating events, but these events occurred slowly over millions of years of geological time periods. What will our time period be today if we continue down our current path? Five hundred years, a thousand years, possibly more if we are lucky.

Earth's sixth mass extinction is currently accelerating and it is the only one in the planets history to be caused by human activity. "Drastically increased rates of species extinctions and declining abundance of many animals and plant populations are well documented, yet some deny that these trends amount to a future elimination." said lead researcher Robert Cowie. Writing in the journal Biological Review, he and his colleagues estimated that between 7.5% and 13% of today's Earth's two million known species may have already been lost. Some critics of the manmade "biological annihilation" of wildlife say this is merely a new natural trend with humans just playing the dominate role in Earths evolutionary history. What of man's future evolution?

Can we count on technology to bail us out of this dilemma? New companies are starting to address the challenge of removing CO2 from the atmosphere, but the outlook for a near term miracle solution is not good. A new large experimental operation in Iceland does remove 4000 tons of CO2 per year, but it would take 9 million of these plants all operated by non-carbon energy sources to remove the carbon emissions that the world generates in just one year. And today the process is very expensive, costing between $600 to $800 dollars for each ton of carbon removed. Iceland has a huge advantage, most of its electricity is generated by non-polluting geo-thermal installations.

It is tempting to assume that global warming is a slow-moving crisis and that we can rely on technology to put the CO2 genie back in the bottle. We must shake off this complacency. We have not invented a practical and safe method for removing more than a trivial fraction of the existing CO2 in our atmosphere and the amount that will be added in the future; planting trees will not be the long-term answer.

Years ago in 2006, James Lovelock a scientist of renowned reputation and the inventor of the microwave oven told an audience that the Earth was heating and could increase as much as 8 degrees

Celsius in the future. He said this would make large parts of the earth's surface uninhabitable if corrective efforts were not taken; threatening billions of people's lives. He said that such a warm earth may be only able to support less than a tenth of its six billion people at that time. "We are not all doomed, an awful lot of people will die but I don't see the species dying out on earth, but at this increase in temperature the earth couldn't support much over 500 million people." Others have suggested that one billion survivors would be more realistic.

In March of 2009 at a Copenhagen scientist conference, Professor Hands Schellnhuber, then director of the Potsdam Institute, and one of Europe's most distinguished climate scientists, told his audience, "At last we have accomplished something, we have estimated that the carrying capacity of the planet at 4C warming would be somewhere below one billion people."

Professor Kevin Anderson of Lyndall Center for Climate Change believes only about 10% of mankind would survive if global temperatures rise by 4 degrees Celsius. Anderson who advises governments on climate change said the consequences would be "terrifying for humanity, it's a matter of life or death. All humans will not go extinct, few people with the right source of resources and living in the right locations will survive. But I think it is highly unlikely that we wouldn't have massive deaths at 4C."

There are those who argued that these projections go too far and are extremely pessimistic.

Let's examine what a 4C increase in the average world's temperature might look like.

If the present trends continue, we may well reach or exceed 4C by the end of this century. Today Earth appears to be heading towards 1.5C increase by 2030 and 2C by 2050 and if the feed-backs kick in; 4C, 30 to 50 years later.

It may take several centuries or even a millennium to melt completely the ice caps at the poles but sea levels could still rise up to 2 to 3 meters (6-9ft) by 2100. And eventually with a 4C rise the earth would have no large ice sheets and the Himalayas would lose their snow and ice cover. This would finally rise sea levels to 70 meters (210ft). These figures seem extreme but the long history of the Earth, as measured by fossil record and ice cores taken in the Antarctic, tells us that for the amount of CO_2 that will be present in the atmosphere it will create this kind of havoc; it has happened before in Earth's history that contained similar amounts of CO_2. Almost two billion people in Asia depend on the water supplied by the snow melt of the Himalayan mountains.

There would be few coral reefs left. Ocean ecosystems and food chains would collapse due to the increase in the ocean temperatures. Today the oceans are already warming and we haven't even reached 1.5C. This will accelerate the long slide into Earths sixth extinction. Some species will adjust and survive given enough time but the deterioration of the dying plants and animal life will create even greater emissions of carbon dioxide.

The warming ocean waters will decrease the production of phytoplankton and the oceans' ability to act as a sink by removing carbon dioxide from the atmosphere. The oceans warm upper surface waters do not mix well with cooler, nutrient- rich deep waters below preventing up-welling and shutting down these essential life sustaining nutrients.

Hundreds of billions of tons of carbon in the Artic permafrost, mostly in Siberia, will began melting (they already have), releasing global warming methane and carbon dioxide in huge quantities. The Amazon Forest will continue to shrink and will no longer be a rain forest capturing carbon but a net emitter of this gas.

Extreme drying will persist over more than 30% of the earth's surface. Desertification will become severe in Southern Africa, the Southern Mediterranean, West Asia, the Middle East, Australia and across the south-western U.S. Agriculture will become impossible in these dry subtropics.

A wide belt of high humidity and heat stress will persist across most of Asia, Africa, Australia and the Americas that will render them unhabitable for most of the year. Some of the climate models predict that the desert conditions will stretch from the Sahara through the south and central Europe; drying up rivers including the Rhine and the Danube. In the summer of 2022, both of these rivers were severely stressed and reached their lowest historic levels. Despite the fact that there will be more intense rainfall the hotter soils will increase evaporation and many populations will have difficulty obtaining fresh water.

Food production will decrease dramatically as result of a decline in crop yields. The nutritional content of the food will be lowered; a catastrophic decline in pollinating insects will occur; desertification; monsoons and hurricanes intensities and destruction will increase; chronic fresh water shortages and conditions too hot for human habitation in the ideal food growing regions will occur.

The jet stream will be altered that will affect all of the earths weather patterns including the distribution of the Asian and West Africa monsoons. The further slowing of the Gulf Stream will influence life support systems in Europe. New deserts will be spreading in Italy, Spain, Greece and Turkey. In Switzerland, summer temperatures will hit 48 degrees Celsius, more like Baghdad than Basel. England could have summer temperatures reaching a scorching 45C. It may well occur that large numbers of Europe's population will be forced to drift north.

The oceans will develop more dead zones where life cannot be sustained. Coral reefs, shellfish and life sustaining plankton will be distressed by the rising acidity and algae starving the oceans of oxygen and its food chain of small food fish. Without such prey the larger sea life will decline rapidly.

So, did the one billion survivor prediction go too far? Not at all, there are serious research and knowledgeable voices in support of that conclusion.

"One must always remember that civilization is a flimsy cloak, and just outside are hunger, thirst, and cold waiting. They are always there and, in the end, unless man remembers they will always defeat him." Louis La Amour

These are just some of the likelihoods of a world warming an additional 4 degrees Celsius. They are projections of possibilities based on numerous scientific studies and climate models. They are not cast in stone. When and whether they will occur depends on the actions the world takes today to dramatically slow down the production and emissions of the greenhouse gases we are continuing to release daily into our increasingly warming world. We would not be the first civilization to collapse but we may well be the last.

A new study reveals that some birds in the American Midwest are now laying their eggs about a month earlier than they did a century ago, with a steadily warming climate as the cause.

Led by Chicago's Field Museum, a team compared century-old eggs preserved in the museum's unique collection with recent avian observations.

Each egg is accompanied by a label, noting the kind of bird and precisely where and when it was collected. A third of the species studied around Chicago now lay their eggs about 25 days earlier than they did a century ago. This gradual shift to an earlier spring has resulted in large impacts on animal and plant life cycles. Scientist believe this may be one reason for the steep decreases in bird populations since the 1970s. Grassland and shore birds have shown sharp declines and more than half of the populations of U.S. bird species are revealing reproduction failures according to The American Bird Conservancy.

A record number of Florida's protected manatees died during 2021; the 1,101 deaths more than double the five-year average. Most of the deaths occurred in the Indian River Lagoon where pollution caused algae blooms killed thousands of acres of seagrass, the manatee's main food source. They are being fed lettuce in order to sustain the population.

Unless climate change is curbed, Earth's oceans could see mass extinctions of marine life unlike anything we have seen for millions of years, according to a new study. "If carbon dioxide emissions increase unchecked over the next century, it would lead to severe warmings driven extinctions as great as the mass extinctions in Earth's past," the study's lead author Justin Penn of Princeton University said. The study revealed that the climate driven ocean warming and the depletion of its oxygen would be the primary reasons for the potential mass extinctions. Human impacts such as habit destruction, overfishing, plastic and costal pollution will also be a factor.

As ocean's temperatures increase, they produce less oxygen and the marine life rapidly decreases. These extinctions could occur in the next 100-300 years. According to the study, if emissions are not severely decreased all marine ecosystems are likely to equal the severity of the Permian extinction that occurred 250 million years ago and led to the demise of two-thirds of all marine animals.

Tropical waters would experience the greatest loss of biodiversity, while the polar species have the highest risk of going extinct. Even though tropical oceans would be expected to lose the most species, many species will migrate to higher latitudes and cooler waters where they have a chance to survive. Polar species would be more likely to go extinct as the waters warm with less ice is available, they can't adjust, they have no place to go too.

Polar bears in Canada's Western Hudson Bay are continuing to die in increasing numbers a latest government survey has found. An ariel survey in 2020 around the town of Churchill, known as "the Polar Bear capital of the world," was made where the bears gather in large numbers. The recent survey showed that there were 618 bears as compared to a 2016 survey that indicated 842 animals; since the 1980s the population has fallen nearly 50%. The diminishing sea ice, due to the Arctic's warming has left the bears less ice on which to live, hunt and reproduce. "These are the types of bears we have always expected to be affected by changes in the environment," said lead investigator Stephen Atkinson who has studied the polar bears for more than thirty years.

The study's news was not all harmful, it found that by reversing greenhouse emissions the risk of extinction could be reduced by more than 70%. "The amount of extinction that we found depends entirely on how much carbon dioxide we emit today and in the future."

Tree Extinction: An international group of scientists issued a "warning to humanity" over the prospect of losing about a third of the world's tree species. Writing in The State of the Worlds Trees report, it says that more than 100 known tree species have already become extinct, with billions of individual trees being lost each year to fire, pests, disease, invasive species, drought, climate change and industrial scale logging and deforestation. The report states that further losses will lead to major biodiversity disruptions in a world where forest provide homes to about 75% of all bird species, 68 % of mammal species and as many as ten million species of invertebrates. If we don't protect the trees there is no way possible to protect all of the life that calls them home.

Amazon Rain Forest: The Amazon rainforest is the largest in the world. This forest covers an area totaling over 4 million square miles; it is the home to more than 10% of the world's biodiversity and its rivers are responsible for about 15% of the world's total river flows into the ocean. It is one of the most important land ecosystems on earth yet it is shrinking rapidly. Human deforestation is the major cause. According to a 2019 report, 17% had already been lost and since then the destruction has increased each year. It is projected that by 2030, 27% of the Amazon will be gone if the current deforestation rates continue. This loss of forest cover is primarily a result of illegal logging to clear land for soy bean farming and cattle ranches. There does not seem to be any plans to stop or slow down this development. Brazil's current president has encouraged both agriculture and mining in the Amazon and has ignored the illegal burning and forest removal since he entered office. Hopefully a new administration will surface that has more concern and will take the necessary measures to reduce and eliminate the depletion of this vital source of carbon capture.

So why is the Amazon Rain Forest so important and why is the clearing of its land such an issue? First, it is the home of over 24 million people including hundreds of thousands of Indigenous peoples that have lived there for thousands of years. It contains 40,000 species of plants, 400 species of mammals, 1300 different kinds of birds and unknown insect population in the millions. It also a major carbon sink, storing over 150 billion metric tons of carbon, keeping it from entering the atmosphere. This is more than a third of all the carbon stored in forest worldwide. Deforestation has caused a decrease in the amount of carbon it can store with some reports stating that up to 20% of the forest is emitting more carbon dioxide today than it is absorbing. Further forest loss will lead to a tipping point when the rainforest cannot recover. This will have a devastating impact on the global climate.

Overfishing: Large areas of the North Sea and the Atlantic Ocean have been overfished for many years. Halibut populations were once so large off the north-east coast of North America that it was not unusual for a single fishing boat to harvest over 20,00lbs. of halibut on a single trip. Halibut numbers have now declined to the point that they are almost extinct in that area. The cod fishery that has existed for a hundred and fifty years off the Georges Banks and Grand Banks had populations that were so dense with cod, some weighing over a hundred pounds, they were thought to be inexhaustible. By 1960 the number of spawning cod in the North Atlantic had fallen to 1.6 million tons. By 1990 it had fallen to 22,000 tons. It was feared that the cod might have been lost forever. In 1992 cod fishing was closed and fortunately since then the species has begun to recover.

Alaska cancels snow crab and red crab season: The Alaskan Department Fish and Game has canceled the fall Bristol Bay red king crab harvest for the first time and are holding off the winter harvest of snow crabs. These decisions came as a result of the sharp declines in populations. Data has shown that the Eastern

Bearing Sea snow crab population has decline 92% from 2018 to 2021. The last snow crab harvest was 5.6 million pounds, the smallest harvest in more than forty years. The numbers shrank after the Bearing Sea warming caused stress on the populations due to the hotter waters. The fall red king crab harvest was canceled because of the low numbers of female crabs that indicate the health of the population. As recently as 2016 the Bearing Sea crab harvest grossed 280 million dollars.

On the Nekasa River in Canada's British Columbia, researchers at the Simon Frazer University discovered more than 65,000 dead pink salmon found in a dried-up waterway before they were able to spawn. The researchers stated that this population of salmon will take up to six generations to recover even if there is no more die-offs events. It was an utter unprecedented devastation of a valuable fishery. As the earth continues to overheat, droughts in British Columbia are becoming more frequent with abnormally low amounts of rain, less runoff and higher water temperatures causing greater evaporation. These conditions triggered many parts of some rivers to completely dry up. Salmon are large part of Canada's seven-billion-dollar fishing industry and provide a staple food source for the native Indian communities throughout the country.

"Man must obtain his knowledge and chose his actions by the process of thinking which nature will not force him to perform. Man has the power to act as his own destroyer and that is the way it has been throughout most of history." Ann Rand

Dams

Dr. Charles Berkey, one of the world's foremost hydrologists, Professor of Geology at Columbia University in 1946 gave a dire prediction of the death of many of man's most impressive monuments: Dams. He stated there is no permanent cure for their eventual destruction, the cause: silt.

When you look at the Grand Canyon did you ever wonder where did all that soil go when it was washed and eroded by rain and wind over millions of years to form the beautiful landscapes we all admire? The Grand Canyon is a fairly young canyon in geological time, approximately ten-million-years-old. At its start, the Colorado River ran across a high desert valley and down through eons of time cut its path deeper and deeper as it worked its way south to the Gulf of California. It is still cutting and eroding. But where did the thousands of billions of tons of soil go? We are farming it now in the Blyth, Yuma and Imperial Valleys in California and Arizona and the Mexicali Valley in Mexico. These vast rich agricultures areas are the heritage the river has left us. When the Colorado River was free flowing without being impeded by dams it carried vast amounts of soil over millions of years that washed down and formed the beautiful colored spires we see today. As the river was released from the canyon it spread out over hundreds of square miles, depositing layer upon layer of silt creating one of the most productive and fertile agriculture areas in the world. It left a delta at the upper end of the Sea of Cortez of hundreds of square miles of tidelands and mud flats that can be seen clearly today from the air.

Before Hoover Dam was built on the Colorado River and before Elephant Dam was constructed on the Rio Grande River both rivers ran chocolate brown in the spring and any time a cloudburst occurred in their drainage area. The Colorado, one of the siltiest rivers on earth carried sediment loads close to those of the much larger Mississippi River. Now the water pouring from Hoover Dam's penstocks and spillways are flowing a beautiful blue-green, colored only by minerals and algae. It has more people, more industry and a more significant economy dependent on its waters than any comparable river in the world. It provides over half of the water for the greater Los Angeles area, San Diego and Phoenix and grows much of America's fresh winter vegetables.

Each year, millions of cubic yards of silt are coming to a dead end in the lakes behind the dams. It is no longer being dispersed, it has nowhere else to go. All dams are carefully constructed to resist earthquakes, landslides and flooding, but their ultimate weakness is silt. Every reservoir eventually silts up-it's only a matter of time.

With the exception of the Nile River in Egypt, the Colorado is the only other river where so many people are so helplessly dependent on one rivers flow. In terms of annual flow, the Colorado isn't a big river. In the U.S. it does not rank among the top twenty-five rivers. When the river was free flowing the silt would start to settle about two hundred miles above the Gulf of California where its flow rate slowed down. There was so mush silt that it raised the entire river bed foot by foot, year after year until it created a new channel carrying the silt with it. Over the ages it changed its course during many occasions on its way to the Gulf. It would fill the Salton Sea, the largest inland body of water in California, and then return to its main channel while the Sea would dry up and become a dry lake bed. Today the Sea is slowly drying up again and will eventually return to its ancient dry past.

Some examples:

The Sanmenxia reservoir in China, an extreme example, was completed in 1960 and was already out of use by 1964, it had silted up completely in just four years.

The Tehri Dam in India, the 6th. highest dam in the world saw its projected life reduced from one hundred years to thirty years. In the Dominican Republic, the eighty -thousand kilowatt Tavera Hydroelectric Project, the country's largest, completed in 1973; by 1984 silt behind the dam had reached a depth of eighteen meters (54 feet) and the storage capacity had been reduced by 40%.

A report from the U.S. Geological Survey and Bureau of Reclamation shows that Lake Powell has lost 1,833,000-acre feet or 6.79% of its storage capacity. The reduction is attributed to the buildup of sediment (silt) from the Colorado and San Juan Rivers. This has happened in just 60 years since the dam was constructed. Today as the water level continues to fall; Lake Powell is down 43 feet from a year ago and is at 23.9% of its full pool capacity, this silt buildup means that there is even less water available and is creating serious concerns for its ability to generate hydropower.

Deforestation, fires, erosion and major storms are the major causes of reservoir siltation. They destroy the landscape that controls erosive runoff, filling the streams and rivers with silt and ash that is fed into the reservoirs. Erosive forces are hard at work in the watersheds of the Missouri River, the Colorado, the Rio Grande, the Platte, the Arkansas, the Brazos, the Pico, the Willamette and the Gila: all rivers on which there are dozens of dams.

Examples of the reduced capacity of dams in this country due to silting.

Black Butte Reservoir, Calif. Capacity in 1963- 160,000-acre feet (af), capacity in 1973-147,754af.

Concha's Reservoir, New Mexico- capacity in 1939, 601,112af, capacity in 1970, 528,951af

Alamogordo Reservoir, New Mexico, capacity in 1936, 156,750af, capacity in 1964 110,655af

Lake Waco, Texas capacity in 1930, 39,378af capacity in 1964 15,427af

Lake Mead, Arizona capacity in 1936, 32,470,000af, capacity in 1970- 30,755,000af

San Calos Reservoir, Arizona capacity in 1928, 1,266,837af, capacity 1966 -1,170,000af

In just 35 years Lake Mead behind Hoover Dam was filled with more acre feet of silt then 98% of the reservoirs in the U.S. are filling with acre feet of water. This rate of silting has slowed at Lake Mead since 1963 because the silt is now building up behind Flaming Gorge, Blue Mesa and Glen Canyon dams after they were constructed. What happens when they are silted? All of the above reduced reservoir capacities are for a time span of only 30-60 years. Silting will continue, there are few ways to reduce its impact. What will the capacity reductions be in the future and how soon will we have a lot of spectacular new water falls spilling over the crest of these large dams?

Again, we must stretch out our time frame. If we intend to still be around in the next 100-300 years how will our water storage and distribution systems cope with this loss of water storage capacity and electrical generation? Even if only 30% of the dams silt up in 150 years that will still present a tremendous challenge. Populations and demand for water and power will surely increase. We had better start thinking and planning now, how we are going to increase, at any cost, the conversion of sea water to fresh water on a massive scale, and how to recover the lost energy and storage capacity. These sea water conversion plants would require enormous amounts of energy as these systems are not energy efficient. There are not too many other alternatives. Even by importing water from distant wet areas of the country, say the Mississippi River or the Columbia River, where are you going to store it; behind silted up dams? The Mississippi River this year has its own problems with drought and low water flows. One of its tributaries, the Platte River in Nebraska, nearly dried up in 2022.

Our ground water has already been seriously depleted in many areas of the country due to excessive over pumping in the last fifty years since the development of the centrifugal pump. Can we continue pumping at the same rate for another hundred years? Do we build more dams and where? Most all of our rivers have already been saturated with dams and there are few if any adequate sites left, you just can't place a dam anywhere. But even the sea water option presents mammoth problems with the rising sea levels due to climate change and the lack of distribution systems. How do you get the converted sea water to the productive farmlands that are far inland from the oceans? Presently we are pumping water from the desert dams on the Colorado

River to storage dams near the coasts, in the future we may have to reverse that and pump it to the desert and other inland locations. But then again where do you store it? In the history of the world entire civilizations have collapsed and disappeared because of lack of available fresh water. Are we destined to repeat history in some locations of this country in the next two hundred years or sooner?

The cities of Los Angeles, San Diego, Phoenix, El Centro, Yuma and many other communities could not exist in their present state without imported water and what is the source of much of that imported water? The desert: A single river, the Colorado. And now in the face of increasing and extended 100-degree daily temperatures, reduced snowpack, early runoff and increased evaporation how secure is that source today and in the future?

From an article in USA Today: Another concern that must be considered is the risk of dam failures due to the increasing influence of climate change, with increasing amounts of river swelling rainfalls. Many dams in the country are in poor condition, nearly 3000 dams have been declared to be inadequate and in need of repairs. There are 7000 other dams located in areas that already are experiencing increasing frequent rainfall events with little or no public information about their condition; many are 75 years old or older. In the recent infrastructure bill passed by Congress, 3 billion dollars was included for the repair of country's aging dams. The Association of State Dam Safety

Officials estimated the cost to repair the dams not owned by the federal government to be 75 billion dollars.

The failures are already happening. In 2020 a heavy rainstorm destroyed the Edenville Dam in central Michigan, causing 11,000 people to evacuate. The owners knew for years that the floodgates were in poor condition and couldn't be opened all the way. In 2019 a severe storm with high winds and heavy rains sent a torrent of water and river ice into Nebraska's Spencer Dam; the dam was breached. It was known the dam had problems with ice jamming the floodgates in the past. In 2015 an offshore hurricane fueled a rainstorm that caused the failure of 36 dams. The National Ocean and Atmospheric Administration said those failures contributed to the flooding that killed 19 people and caused damage to at least $1.5 billion worth of property. More than 200 dams have failed due to downpours, floods and other weather since 2000. Many were smaller dams but the destruction due to flooding destroyed vital infrastructure.

John France an engineer that has worked on the repair of several major dam failures said, "The methods we have been using have been based on past performance. Now the science is telling us, with climate change, that is probably not a valid approach." The climate data is clear: Intense storms are occurring with increasing frequency. Just in the space of four weeks during the summer of 2022 six storms classified as having a chance of one in a thousand occurring in any given year occurred in the U.S. In Texas 15 inches of rain fell in 24 hours that flooded Dallas eastern suburbs. There were two such storms in 2021.

Rainfall projections calculated at the University of Wisconsin predict that as increased temperatures rise the trend of increasing heavy rainfall, flooding events east of the Rocky Mountains will be more frequent. For more than 5000-high and significant hazard dams, what was once a 25-year storm in 1995 is now expected to become a 15 year-storm by 2055. Inspectors have classified the conditions of 940 of those dams as being in poor or unsatisfactory condition. What was once a 25-year storm in Oklahoma City; 6" of rain in 24 hours, has become 20% more likely in any given year according to the University of Wisconsin data.

Migration and Human Impact

Research has shown that human migrations throughout the world has, due to the effects of climate change or other causes, forced people to move in growing numbers to where there are available resources of water, food or a safer political environment that is more stable with greater opportunities; over the last twenty years.

According to a recent United Nations report weather disasters linked to climate change have pushed roughly 21.5 million people in countries already struggling with conflicts to move each year since 2010. As the earth continues to warm there will be more people experiencing calamities like flooding, landslides, hurricanes, excessive heat, water and food shortages and social unrest that will cause increased instances of large human migration. It will become impossible to control or slow down. As an example, a third of Pakistan today is flooded with millions of people displaced from their homes; where can these people go?

If global warming only reaches 2 degrees Celsius increase this century, half of the world's population could still suffer from climate hazards. Millions of businesses and homeowners in the United States will be compelled to move because of sea level rise, consistent flooding, lack of fresh water, droughts and other warming impacts. In fact, this migration has already begun. The Urban Institute estimates more than 1.2 million Americans have moved and left their homes for other locations due to climate related reasons.

This is another aspect of climate change that has not been fully considered as a serious threat to mankind. Until this country and the rest of the word get serious about the challenge of the reducing the carbon that is daily flowing into the atmosphere it is only going to get worse with massive number of desperate people flooding the roads, highways and seas seeking a better life. The impact on a country's borders, infrastructure, housing, food and water supplies and political pressures when trying to control and accommodate the influx of thousands of migrants will be severe, sometimes leading to strife and conflict. We are seeing that today in our country and in other countries where conditions are forcing relocation.

Central American populations including Guatemala, El Salvador, Honduras, and Nicaragua for years have been facing life-threatening levels of food insecurity, blistering drought, intensifying storms and internal strife of wars and government upheavals. The frequency, length and intensity of these conditions will continue to increase as temperatures continue to warm to extreme levels near the equator where they will be the most dangerous, causing greater suffering and the necessity for people to relocate.

According to the World Food Program, about 828 million people around the world go to bed hungry every night with 50 million people on the brink of starvation. The number of people afflicted by hunger has grown by 150 million since before the COVID-19 pandemic, according to a report on The State of Food Security and Nutrition.

The Web: The climate calamity is intensifying the U.S. border crisis. In the face of political turmoil, a magnitude 7.2 earthquake that killed over a thousand people, severe tropical storms and intense heating, Haitians are fleeing their country. More than 10,000 migrants from Haiti converged on the U.S. border at Del Rio, Texas recently and an additional 30,000 more may also be seeking to travel north. Their experiences are similar to those migrants from Central and South America where political and economic instability, hunger and heating due to climate change are threatening their livelihoods.

This present U.S. border crisis paints a picture of how natural disasters, food and water shortages, political upheaval and weather changes can push people to leave their homes, even if it means risking their lives. It is important for the world to understand how these people, not only on our borders, have become particular vulnerable and personally exposed to the heating planet. It will become a floodtide for survival, people are not going to halt their movement seeking a better life when they have no other choice, and we are not going to be able to stop them.

When they are not able to obtain the basic resources of food, clean water and are continually exposed to landslides, flooding, wars and hurricanes and political pressures that are happening more frequently, they migrate hoping for a better future, they have no other choice. Food Crisis: More than 43% of the Afghanistan people are living on only one meal a day. Ninety two percent of the population were experiencing insufficient food consumption in June of 2022. In the last year more than 6 million Afghanistan's have been driven from their homes; 780,000 have fled and are living in Iran and an additional one million have escaped to Pakistan.

Recently the World Economic Forum released the results of a climate change survey. It found that approximately 56% of people across 34 countries said that climate change has already impacted the areas where they live. In the U.S., about 48% of those surveyed said that the climate has had a severe effect on their lives. The widespread flooding in Kentucky and other areas, and those living in the West experiencing severe drought and wildfires were a factor. About a third of Americans believe it is likely that they will be displaced from their homes in the next 25 years.

The recent Florida hurricane Ian should be a wake-up call, not only for Florida but the rest of the nation. Ian now ranks among the top 30 deadliest storms in U.S. history. It is also the 11th. storm in the past 22 years to claim more than 50 lives. The risk of these storms will increase because of the warming ocean, sea level rise and increasing population density. Amber Silver, a disaster researcher and assistant professor at the University Albany, New York said, "When you have increasing populations in regions that are highly vulnerable to climate change and sea level rise, we will continue to see these events with large death tolls and economic losses." There are more people moving into costal locations and exposing themselves to vulnerable locations. Florida is the prime example. Since 1992 its population has increased 60% and the number of homes has doubled. Previous hurricanes had conditioned people to fear the wind but as many as 90% of all hurricane deaths can be attributed to storm surge and heavy rainfall. Ian had storm surges of 12 to 18 feet that pushed the sea far inland. State medical examiners attributed more than half of Ian's known deaths to drowning.

It is always difficult for people to evacuate and leave their homes, even knowing the potential risk. For some because of age, financial conditions, lack of transportation or misinformation it may be impossible to leave. Today the lack of truth in our society may steer some people to ignore or not believe the warnings they receive from reliable sources. They think what they are being warned about is media lies and misinformation or scare tactics. This has become serious problem for our times when people have been indoctrinated to the lie and the lack of a universal belief in the truth. Unfortunately, many people have died because of this attitude.

Where will we end up living?

In large portions of the world local conditions will become too warm and there will be no way to adapt. The people will be forced to move to survive the changing situations. Over the remaining portions of this century and beyond, increasing temperatures will make large areas of the planet uninhabitable. People will need to leave the coastal areas, the Tropics, Africa, Central and South America and some island nations where it will be no longer possible to farm or even work outside because of the heat or rising seas.

According to the United Nations the number of people globally fleeing their homes and searching for a new life has doubled over the past ten years. As the planet continues to warm and more people are displaced large migrations will continue. Where will they be able to go to find adequate fresh water, safe temperatures, agricultural land that that is productive and able to sustain increasing migrant populations? Unfortunately, there will be no place on earth that will be completely free of the future warming effects. Everyone on the globe will experience some form of the consequences of the changing climate. There will be no truly safe places but there will many areas where people can adapt and live; mostly in the Northern Hemisphere. We had better start thinking about where those livable areas are and plan for these inevitable changes.

If they cannot move to sustainable areas, it is estimated that one third of the global population will experience average temperatures like those that are found today in the warmest spots on earth: Death Valley and the Sahara Desert. Northern parts of the globe with higher elevations and distant from the equator, that have adequate fresh water and cooler temperatures will be a magnet for migration. These inland areas will still have increases in average temperatures but they will be far lower than those in the equatorial tropic regions. Inland lake areas like the Great Lakes and the many lake areas in Canada will have a cooling effect and will have vast increases in populations. Cities in the northern areas will attract people escaping the severe condition in the south. Beach front properties will not be as desirable as they have been in the past.

It should be remembered that many places on the globe will be uncomfortable by 2050 and beyond. Planning should start now to project where people are going to live. There are many options: Northern United States, Canada, Alaska, Russia, Iceland, Scotland and the Nordic nations. What will be the political and social ramifications of massive numbers of people, many of them poor, moving into the more northern wealthier countries? How will these countries deal with these displaced millions? The movement of people we are seeing today are only a small percentage of what the world will be realizing over the next 100-200 years.

Coastal cities like New York, San Francisco, Seattle, San Diego and their populations will face enormous threats but they may be thought to be too important to fail. But therein lies the dilemma; how are they going to be protected from sea levels rise of six to ten feet by the end of the century and the continuing rise in the decades and century's following? Surely, large numbers of the populations of these and other coastal cities around the world will have to move to cooler havens to escape the unlivable heating and rising oceans. As previously stated, many of the northern areas will be cooler sanctuaries. Alaska will become one of the most desirable locations and one that has the lowest climate events of the country. By 2047 Alaska could have average monthly temperatures equivalent to those of Florida. The Nordic nations also have a low vulnerability to climate change and will experience increased immigration.

Researchers have estimated the declining growth of many of the southern countries. India's gross domestic product has declined 31 % due to global warming; Nigeria's is down 29%; Indonesia's by 27%; and Brazil's by 25%. Those four countries include a quarter of the world's population. Where are they going to go when their area's become unbearable? Russia will profit from the coming climate warming. According to the U.S. Intelligence Council, Russia "has the potential to be the biggest gainer from the increasing warmer weather." It is the largest wheat producer in the world and that will only increase as the climate warms. By 2080 more than half of Siberia's permafrost may have disappeared promoting longer growing seasons and opening huge areas of new lands for cultivation. But on the downside, they will have serious and very expensive problems as the permafrost melts under many northern settlements and cities. The frozen soil that formally supported the foundations of buildings, pipe lines, roads, rail lines and other infrastructure will subside under these structures and they will very expensive to replace or repair. In some areas this is already happening.

If people have the opportunity to relocate, they will move to safer places with stable governments, job availability and natural resources. In some areas entire new cities will have to be constructed. So, all is not lost, mankind will survive. It will be a new and difficult experience for hundreds of millions of people and will require rapid adaptation to stay ahead of the continuing changes of the warmer world.

Bird migration: Climate change can add to the threats for many migratory bird species. It can upset the relationship between migration timing and the availability of their food resources. Hungry birds after a long migration, may arrive in their typical feeding areas to find that the peak of the abundance of insects they feed on has already passed.

Heat Waves, Heat Domes & Heat Islands

Heat Islands; What are they and how do they occur? Large buildings tightly packed within a metropolitan area; concrete and asphalt roadways, parking areas and sidewalks, dark roof areas absorb and trap and hold the heat. The large buildings prevent air movement that would have a cooling effect. The heat that is absorbed is slow to cool and if the air temperature is hot the heat islands add several degrees of additional heating to the discomfort of the residents and even at night there is little cooling.

Heat Domes: A heat dome occurs when a persistent region of high-pressure traps heat over an area. The heat dome can stretch over several states and linger for days or weeks, leaving the people, crops and animals to suffer through stagnate hot air that can feel like an oven.

The climate crisis has increased the intensity of heat waves, heat islands and heat domes in populated areas. The number of heat days per year and the length of these heat waves has increased. The average length of a heat wave season in 50 large cities is now about 70 days compared to 20 days in the 1960s. In less than one lifetime the heat wave duration has tripled. This has multiplied the heat island effect for millions of people.

An unprecedented and early heat wave in June, 2022 has made life almost unbearable across Pakistan and India. The heat started huge landfill fires and caused power blackouts as more than 1.5 billion people try to keep themselves and their food cool with air conditioning and refrigerators which are in short supply. Electricity supplies are unavailable 18 hours each day in some areas. Crops are perishing in heat and drought, worsening already acute food shortages. And now in Pakistan, after all that heat, one third of the country is under water due to extensive rain and glacier melt and millions of people are displaced and made homeless.

Temperatures soared above 110 degrees and, in some areas, reached as high as 115 degrees. People had little relief for weeks with the effect particularity hard for millions of outdoor workers. The heat raised temperatures 15 degrees above normal across much of India as reported by NASA's Earth Observatory. March 2022 was the hottest March India has ever recorded since record-keeping began 120 years ago. Heat waves have been found to be the deadliest of the extreme weather events and they are the type of events that are increasingly occurring in a warming world due to climate change.

The summer of 2022 drought and heat in China were the most severe ever recorded. The nearly stationary heat dome has lasted longer than any other in history and has forced factories to close threatening to upset global supply chains. The heat has evaporated reservoirs and rivers, disrupting hydroelectric turbines that provide power to millions of Chinese. China's autumn harvest may now be lost, which could worsen the already acute global food crisis. "There is nothing in the world's climate history that would be even minimally comparable to what is happening in China today," said climatologist Maximiliano Herrera.

The unprecedented heat that has baked the Northern Hemisphere the summer of 2022 has caused railroad tracks to buckle, forced airport runways to be shut down and damaged roadways. These transportation difficulties, due to the heat, have affected areas from Africa to China and even a busy highway in Cambridge England. Experts warn that today's modern systems of transport infrastructure were designed and built for much cooler conditions.

Anywhere it is humid and hot the heat effect can feel worse. In Florida researchers have tested the impact of heat islands with high humidity. On 58th Street in West Palm Beach at a location without shade trees, the temperature reached 93.9 degrees near noon on July 22 with a relative humidity of 58% according to the measurements taken. The combined heat and the high humidity resulted in a human discomfort reading of 106 degrees in this heat island.

The characteristics of the neighborhood people live in contribute greatly to effects of increased heat. The consequence of the effects on people's health during heat islands warming is more severe on many minority populations. In numerous places their neighborhoods are massed with close housing that provide very little vegetation; lawns and trees to provide shade and cooling. With few or no parks or tree shade and often bordered by industrial areas, highways and other heat retention structures there is little relief. When new developments take out greenery and open spaces and fill them with buildings, highways, parking areas and other structures they help absorb and amplify the heat in the neighborhood.

From 2005 to 2015, the number of temperature related emergency room visits in the U.S. increased by 67% for colored people, 63 % for Hispanic's, 53% for Asian's compared to 27% for white people. This is an indication that neighborhoods make a big difference in the heat effects. The more affluent white neighborhoods with their tree lined streets, large green areas, less crowding and fewer exposed areas to collect the suns heat makes a big difference. On average people of color live in areas with higher surface urban heat island intensity then white people in all but six of the 175 largest urbanized areas in the U.S. according to a 2021 study published in the science journal Nature Communications.

The warming climate has been tied to increased mortality around the world. In a large-scale study that investigated heat in 43 countries, including the United States, researchers found that 37% of heat related deaths were attributed to the changing climate.

The following excerpts are taken from an article written by Elizabeth Weise occurring in the USA Today newspaper.

Scientist have been predicting for decades wilder and increasing weather fluctuations resulting from a warming planet. Because of the rapid changes occurring with today's weather, it is becoming increasingly difficult for them to update and predict future events. As an example, the federal rainfall benchmarks used by engineers to plan and design new roads, flood controls, sewage systems and building have become outdated, in some cases by decades, increasing the risk for flooding by the new normal of the changing climate.

During the 2021 heat wave in the Pacific Northwest, it was previously thought impossible that temperatures could reach 121 degrees in that perennially cool and rainy region- until it happened. Unfortunately, hundreds of people died in a heat dome event that occurred in the last week in June due to some of the highest temperatures ever recorded in the area.

In the South West predictions of increased warming and dryer periods caused by a 22-year ongoing megadrought are occurring that is threating the drinking water supplies for many cities. Lakes are drying up; some crops are failing due to reduced water for irrigation and ground water supplies are rapidly being depleted. Snow quantities have become less and early melt and runoff due to increase heating reduce available supplies.

Death Valley has recorded what is thought to be the hottest ever September day anywhere on the planet; 127 degrees, September 1st, 2022.

"There is a great number of locations where we have sophisticated systems that have been used for projecting the "old climate" said Noah Diffenbaugh, a climate scientist at Stanford University. "Updating climate models and projections to prepare for increasingly extreme events built around assumptions of a previous stationary and unchanging climate is very difficult." It is very challenging to design a world that is resilient to climate change when changes are occurring so rapidly. Using the most current numbers, especially when they so often today are dire, is not always the first choice of policy makers and state and federal agencies, said Kurt Schwabe, a professor of environmental economics and water allocation at the University of California, Riverside. "It's often just kind of wishful thinking" Schwabe said, adding that when given a range of possible values, "they often chose a more optimistic one because they want to make things look good for their constituents, "That doesn't work now when every day brings news of temperature records being broken, excessive flooding, massive fires burning and lakes and rivers running dry."

Extreme heat is scorching Europe: In Britain the government issued a first ever "red warning" for extensive heat when the temperature reached a record high of 104 degrees in Southern

England. This area is usually known for moderate summer heat with July highs in the 70s. By late afternoon 29 location in the U.K. had broken their previous heat records. The U.K. national weather forecaster said such highs are now a fact of life in a country ill-prepared for such extremes. He said that such temperatures were virtually impossible without the effects of human driven climate change. The sweltering weather has disrupted travel, health care and schools. Many homes, small business and even public buildings including hospital do not have air conditioning. Ben Clark, a researcher in extreme weather at the University of Oxford, said, "these extreme heat waves will become far more likely in the future, the recent high temperatures were exceptionally rare before significant human emissions of greenhouse gases."

The summer of 2022, Spain recorded 237 deaths that were attributed to high temperatures. Sevilla, Spain has been a hot spot, recording temperatures of 105 degrees for nine straight days; parts of the country could reach 120 degrees. Spain is also fighting several large wildfires including two that have burned over 18,000 acres and caused 3000 people to evacuate.

In France firefighters struggled in high heat to contain a huge wildfire that raced across the Bordeaux region for a weak.

Temperatures in the interior of Portugal were forecast to hit 111 degrees and could reach as high as 120 degrees. More than 160 people have been injured by wildfires and hundreds have been forced to evacuate. "The chance of temperatures this extreme are already ten times higher than they would have been without human activity," said Nikos a climate scientist with the U.K. Met office.

Climate Shift: Changes in the strength of the prevailing high-pressure systems over the Atlantic Ocean has brought parts of Spain and Portugal their driest climate in over a thousand years. Writing in the journal Nature Geoscience U.S., researchers say the Mediterranean area has become dryer as the high pressure expanded dramatically during the 20th century in step with global warming. This expansion of the Bermuda-Azores high pressure came as the western U.S. also developed the potential for a worsening "megadrought" that threatens cities such as Los Angeles, Phoenix, San Diego and Los Vegas with critical water shortages.

Marine heat waves have brought excessive ocean heating to the Mediterranean that could wipe out several marine species. The sea between Spain and Italy has been up to 9 degrees Fahrenheit above normal during the summer, and it is feared the warmth has already devastated some ecosystems. In the past several years, less severe warming resulted in mass die-off's of marine life. The relentless heat has also damaged crops, including the prize olive trees of Spain and Italy.

Swiss scientists say the freezing point in the Alps has risen to a record altitude above the mountains this summer as Europe bakes in unpresented heat. Weather balloons had to rise to 17,008 feet above sea level to reach the freezing altitude. This was 230 feet higher than had previously been recorded in 1995. Glaciologist Mattias Huss says the accelerating heating has caused glaciers to melt faster than ever. Wild species accustomed to cold climates are migrating higher and are reaching a point where they have no-where else to go.

More frequent and intensive sand storms have occurred across parts of the Middle East and recently even in the city of Phoenix Arizona due to the drying out of the earth caused by global heating. A single storm can last for days and effect many adjoining countries. Storms this spring and summer have filled area hospitals with patients suffering from breathing difficulty. Schools and businesses have been forced to close many times because of the blowing sand and earth. It is believed that the warmer climate with changing weather patterns and poor agriculture management are turning large areas into sand and deserts.

There are conditions that are happening around the world today that are becoming harmful faster than climate scientist thought possible. They are having problems with the severity and how fast it is changing. When Texas officials met with the National Weather Service in 2018 to update their decades-old benchmarks they found that the likelihood of a 100-year storm had changed to 25 years in Huston.

People are already paying the price for living with outdated weather projections in the West. One example is the water allocation from the Colorado River. A legal agreement among seven western states set water rights for each state and the amount of water each state was to receive from the river annually based on the excessive huge river flows measured during a wet period in the early 1900s. "It has never had that much water since, so the possibility of them having that much water, on a yearly basis, in the future is extremely doubtful." Said Gerald Meekl, a senior scientist at the National Center for Atmospheric Research who was part of the team that won the Nobel Prize in 1997 for their work on climate change. This means that each state will receive considerably less water they had planned for to serve their increasing population and growing agriculture needs.

While no one was thinking about climate change a hundred years ago there is no excuse for not having planed for the mega-drought in the West that has lasted for two decades and has become the regions worst in 1200 years. The pace of the drought increases the degree at which groundwater has been depleted due to over pumping and the rate at which ice and snow is melting are some of the events that are changing much faster than scientist can keep up with.

All of the existing and future climate models won't fix the problems the world faces. What will be required is the political determination to immediately lower the carbon dioxide levels that are now being emitting into the atmosphere creating the changing climate, and to prepare for the continuing climate extremes that are certain to be coming in the future. But the big unanswered question is what is the future going to be like? The solution in answering that question is going to depend on the world's political and economic systems. What is the world's political and financial leaders going to do or not do?

Part of the problem climate scientist concede was that the weather changes they began predicting decades ago were so extreme for the times, they could be ignored by those leaders' making decisions on how to plan for the future. Now that future has arrived earlier then some people had expected, but that doesn't make it easier for people to accept the new reality. Efforts must be continued to slow the events contributing to the changing climate and there must be long-term endeavors made to adapt to todays and tomorrow's reality. Plans must be adopted to give more attention to extreme weather events. They can no longer be based on what happened in the past because the future is going to be very, very different.

In India's Gujarat state, wildlife officials have been finding large numbers of dehydrated and exhausted birds as the area remains in the grip of an increasing heat wave. Birds that were still alive were treated with water injections into their mouths and fed vitamins. The heat has also responsible for the deaths of more than two dozen humans across India. Since the country suffered its hottest March in more than a hundred years, the continuing heat has caused water shortages, power cuts and widespread misery, with temperatures well above 100 degrees most every day.

By the year 2100, extreme heat events will make parts of Asia and Africa uninhabitable for up to 600 million people the United Nations and Red Cross just reported. The projected deaths from heat waves will be staggeringly high. The report adds "if little is done to curb emissions" densely populated urban centers in South Asia, the Middle East and North Africa will suffer from "recurring life – threatening heat events" that bring temperatures beyond the human survivability threshold. The past seven years have been the hottest in recorded history. This year, 2022, India and Pakistan suffered continued extreme heat that began in March. It closed schools and cut crop yields reaching a temperature of 117 degrees Fahrenheit. Last year, parts of the Middle East had temperatures of 125F. During a heat wave five years before, a Kuwaiti town topped 129 degrees. These events are happening today, what will the future be in the next fifty or a hundred years?

By the end of this century, one third of the global population could be living in areas with the average temperature is above 84 degrees, mostly in Africa's Sahara region. Extreme heat waves will also make parts of the U.S. less suitable for human habitation by 2070 if global temperatures rise by 2 degrees Celsius. The report concludes by stating: "This is not a problem that humanitarian organizations can solve alone. The urgent priority must be a large and sustained investment today that mitigates climate change.

Without those investments, the world is destined for a future of even larger and deadlier heat disasters."

Spain's earliest and most intense heat wave in 40 years killed hundreds of baby swifts after the hatchlings fled their sweltering nests too soon. The threaten birds were seen littering the streets around the southern cities of Cordoba and Seville. Residents of both cities gathered the dehydrated and starving birds so they could be hopefully nursed back to health.

A new report investigates how dangerously high temperatures could increase over the next 30 years and reveals a grim outlook for much of the nation, especially a vast area of the central U.S. where residents are not acclimated to extreme heat. South Florida is forecast to see the biggest increase in the number of the very hottest days according to a new heat model and assessment by the First Street Foundation, a Brooklyn New York research and technology group. The report suggests that even some of the nation's northernmost counties will not escape the effects of the warming world, said Jeremy Porter, chief research officer.

The foundation looked at average heat index temperatures, what it feels like, based on the temperature and humidity, on the seven hottest days of the year. They considered how climate change could increase the frequency, duration and intensity of dangerously hot days over a 30-year period. The study found that 94% of the nation's counties could experience those days doubling. This coming year more than 8 million residents in 50 counties could experience at least one day with a heat index above 125 degrees. By 2053 that number could grow to more than 105 million people covering more than a third of the country. These findings raise the question as to how people will handle the increased heat; and indicates that communities should be preparing now for these coming conditions.

Today, temperatures feel like 103 degrees or warmer on the seven hottest days of the year in Florida's Miami-Dade County. That could raise to 34 days in 2053, even as Florida continues to have the number one growth rate in the nation.

On Martha's Vineyard in Massachusetts, only three days a year feel like 90 degrees but that could change to twelve days.

The forecast is even more ominous in other parts of the world. For many places close to the equator, by 2100 more than half the year it will be a challenge to work outside. In a worst-case scenario in which emissions continue to increase, by 2100 extremely dangerous conditions will persist in which humans should not be outdoors for any amount of time could be common in countries closer to the equator such as India and sub-Saharan Africa. The study looks at "heat index" which measures the effect of heat on the human body. A "dangerous" heat index is defined by the Weather Service as 103 degrees. An "extremely dangerous" heat index is

124 degrees, which is considered unsafe to humans for any amount of time. It is frightening to think what would happen if 30 to 40 days a year were exceeding the extremely dangerous threshold.

This July was the third -warmest on record in the U.S., with an average temperature of 76.4 degrees, that was 4 degrees above normal, the National Oceanographic and Atmospheric Administration reported. Texas had its warmest July on record, while Oregon just saw its fourth warmest July. The report gives different results depending on location.

According to the report by 2053, 430 counties in 16 states could see the number of days with their current hottest temperatures more than triple. Some other states that will see the largest increase are concentrated in an arc northward from Texas and Louisiana on the Gulf of Mexico, north to Missouri and Illinois, including Kentucky and Tennessee. This region has been described "as an emerging heat belt" because of its risk of exposure to extreme heat index temperatures of more than 125 degrees. This low-lying area east of the Rocky Mountains is region where high humidity interacts with increased temperature. Unlike coastal areas there are no sea breezes to cool the air. It is projected that this area will also have increased rainfall events from the warming of the Gulf of Mexico.

What do these projections mean to people? The increasing potential of heat waves and warmer nights is "really concerning," said Gabriel Fillippelli. Executive director of Indiana University' Environmental Resilience Institute. "People tend to be able to handle one or two days of hot weather but when that stretches to three or four days, human systems start to break down, especially in children, the elderly and those in the lower-income locations. When heat concentrates in urban areas, the temperatures don't cool at night and people get little if any, relief. That can create significant health problems. When overnight temperatures remain high the body doesn't have a chance to recover from the high heat exposure during the day."

Examples of unrelenting heat occurred during the summer of 2022 when high heat alerts were placed in 28 states from California to New York. Temperatures of 5 to 15 degrees above normal were reported across the country. Approximately 100 million people came under an excessive warning or heat advisory the National Weather Service reported. An excessive heat warning is reserved for only the hottest days of the year and is issued by the weather service when temperatures are expected to rise to dangerous levels.

The southern Plains were at the center of the extreme heat meteorologists reported. Dallas, Oklahoma City and Tulsa could all approach 110 degrees. For the first time in history, every one of the Oklahoma's networks of 120 weather stations hit 103 degrees or higher. The heat wave extended into the Northeast where New York City had highs of 90 degrees for five consecutive days.

Boston, Philadelphia and Washington DC also had several days of 90 degrees.

Along with the high temperatures, winds gusting up to 30 mph and drought conditions have created situations in the southern plains requiring red flag fire warnings for central Texas and western and eastern Oklahoma.

Health

"The first wealth is health," said Emerson. It's an old truth but one that bears repeating. The man who wants to work and live life to its fullest cannot take his health for granted. He must take care of himself. "He that has health has everything" goes an old proverb. Guard your health as if it were your most precious possession. It is!

Wildfires are increasing in frequency and intensity in many countries around the world giving off smoke that contains noxious gases, chemicals and particulate matter that carries serious health risks. Wildfire smoke is more toxic than normal air pollution and can linger in the air we breathe for weeks and travel hundreds of miles. It can also contain traces of metals, plastics and other synthetic materials. Laboratory analysis have shown that wildfire smoke causes more inflammation and tissue damage then air pollution. Studies have linked this smoke with higher rates of heart attacks, strokes and increases in emergency room visits for asthma and the effect of weakened immune defenses.

Opinion, by Dr. Neha Pathak: "We are entering a climate that humans did not evolve to withstand. If we continue on our current path of deadly fossil fuel use and the overwhelming temperatures that we are experiencing today, human suffering will only continue to get more intense, more frequent and longer lasting. This means we can expect more health harms, more misery and more deaths as we enter into a climate where humans cannot cope with the increasing temperatures."

Damascus, Saria: Poor access to safe drinking water has spread a cholera outbreak across Saria and other areas of the Middle East. More than 35,000 suspected cases have been reported in Saria in 2022. Cholera is spread through contaminated water, food or sewage that can cause severe intestinal disease and death.

In an annual report by the Lancet Medical Journal experts claim that climate change is the "greatest global threat to health." Air pollution from fossil fuel burning has been found to harm every organ in the body the research shows. According to data from the report, particulate matter from burning coal, diesel and other forms of carbon causes 32,000 deaths in the U.S. during 2020.

As our bodies work hard to maintain an ideal temperature range, the extra heat causes our hearts to pump harder to increase the blood flow to our skin so the heat can escape into the air. We also produce more sweat that helps us cool off. If we do not have time to recover, rehydrate and rest, dehydration can result in kidney damage and increase the hearts work as it tries to protect us by increased circulation. Heat stroke is a life-threatening emergency where our core temperature rises quickly, cells break down and our neve system collapses. Without early medical attention up to 80% of heatstroke cases can be deadly, and survivors can have long term organ damage.

It is not only heat- related deaths that are a concern. Higher temperatures are also linked to higher rates of mental health, more suicides, more violence and lower test scores in children and lower worker productivity. It is not just individual lives at risks, its entire societies.

Active pharmaceutical ingredients (APIs) that are being flushed into the worlds rivers in sewage are now a "global threat to the environment and human health." These contaminants are not removed by the normal sewage treatment process. Scientist at the University of York say they tested water from more than 1,000 sites in more than 100 countries and found many of them polluted with such APIs as epilepsy and diabetes drugs and painkillers. The report also noted that the increase in antibiotics found in rivers could lead to more "superbugs" that developed immunity to their effect. The hot spots for such pollution were in Pakistan, Bolivia, Kenya, Ethiopia. But Madrid Spain, Dallas Texas and Glasgow Scotland, were in the top 20% of the contaminated cities.

The likelihood of extreme epidemics will increase threefold in the coming decades according to a recent study published in the Proceedings of the National Academy of Science. The research found that the probability of a person experiencing a pandemic like COID-19 in one's lifetime is around 38%, in the coming years that frequency could double. The probability of another pandemic is, "almost certainly to increase because of all the environmental changes occurring," said William Pan one of the study's authors and an associate professor of Global Environmental Health at Duke University. Scientists are concerned and are looking closely at diseases caused by germs that spread between people and animals.

Animals can carry viruses and bacteria that humans can encounter through direct contact, or indirectly through soil and water supplies according to the Center for Disease Control and Prevention. Animal contact diseases now account for 60% of all diseases and 75% of new diseases, according to the CDC.

Dr. Aaron Bernstein, director of the climate medical program at the Center for Climate Public Health at Harvard University stated, "that as more animals come in contact with more people and animals that never had contact with other species it increases the opportunity for viruses to be spread animal to animal and then to people." He said "we need to address this mixing by protecting habitats. We must attack climate change and the risk of large - scale livestock production because a lot of the pathogens move from wild animals into livestock and then into people."

Climate Diseases: More than half of the diseases that infect humans from pathogens such as viruses and bacteria have been made worse by the deepening climate emergency, according to a new report. Researchers at the University of Hawaii reviewed more than 70,000 studies of all known diseases that have ever affected humanity, and looked how global heating has affected them.

Writing in Nature Climate Change, they say diseases such as Zika, malaria, dengue and even COVID-19 have been made more severe to humans by climate-related events such as extreme rainfall, floods, droughts, heat waves and wildfires.

They added that altered rainfall patterns have expanded the ranges of disease-carrying pest such as ticks, fleas and mosquitoes which carry malaria, Lyme disease, West Nile virus and other illnesses. The list also included asthma, monkeypox, shellfish poisoning and fungal infections.

"Isn't it ironic and hypercritical that many of the same people who claim their personal choices are being denied because of the requirements for vaccinations are the same people who oppose a woman's right to their personal choice as to what they can do with their own bodies?" Byrum

"The results were truly sobering, "said Erik Franklin, assistant professor at the University's Hawaii Institute of Marine Biology the study's co-author. Climate hazards are bringing pathogens and people closer together, strengthening the pathogens and impairing people Franklin said. "This is a massive vulnerability for human health care systems." The researchers found many unique ways the warming can affect pathogens and human contact. Warming, rainfall and flooding were the three most frequent pathways these pathogens could be transmitted to people. "It just amplifies the key message that the climate crisis is a human health crisis," said Professor Jonathan Patz of the University of Wisconsin.

The study reveals "worrisome glimpses" into the possible consequences of health crisis in the future and points to an "urgent need" to reduce fossil fuel emissions that cause planet warming, Franklin and others said. "There are just too many diseases and pathways for their transmission for us think we can truly adapt to climate change," said co-author Camilo Mora, professor at the University of Hawaii's College of Social Sciences. "It highlights the urgent need to reduce greenhouse gas emissions globally."

This spring the American Lung Association released a report called, State of the Air 2022. "The report shows an unacceptable number of Americans still are living in areas with poor air quality that could impact their health," Harold Wimmer, the associations national president and CEO, stated about the report. The 23rd. annual report also shows there were more days with "very unhealthy and hazardous air quality than ever, even before the two decades of historical reductions." The report states that despite improvement in air quality over the past 50 years, about 137 million Americans continue to live with levels of air pollution.

"There are still far too many days and far too many people breathing polluted air," Paul Billings, the national senior vice president for public policy at the lung association said.

The report deals with the two main types of air pollution: smog (ground level ozone) and soot (particulate matter). Warmer temperatures make ozone more likely to form from auto exhausts, power plant and industrial smokestacks. Ground level ozone is a powerful respiratory irritant whose effects have been linked to lung disease. Soot pollution from wildfires, diesel engines and other sources is deadlier and more of a health hazard then smog, causing more premature deaths and lung cancer, the lung association said. The report based on data from the U.S. Environmental Protection Agency, covered the years 2018, 2019 and 2020.

Florida, Sugar Cane Fields Burning: Florida's state department of agriculture continues to authorize the burning of sugar cane fields and ironically the burning started this year on the same day the state declared the start of "Florida's Climate Week." Over a period of eight months 8000 intentional fires will burn at least 300,000 acres of agriculture land sending large amounts of toxic smoke, ash and soot into the air. Setting sugar cane fields afire to burn off the plants unused outer layer to facilitate harvesting saves the growers millions of dollars each year. This practice has been prohibited in other countries on evidence that the burning contributes to climate change and harms the health of workers and residents in the burn areas; high levels of dangerous particle matter are released. Scientific evidence has clearly shown that the burning harms people's health.

Although there has been long term improvement in the nation's air quality thanks to decades of work to reduce emissions, it has been offset by hotter, drier conditions caused by the change in climate, the lung association reported. Wildfires, increased in intensity by climate change, were responsible for a sharp rise in particle pollution spikes in several states in the western U.S. Overall the report found that 2.1 million more Americans live in counties with unhealthy air compared with last year's report, and exposure to deadly wildfire soot has gotten worse. What's the solution to improved air quality? The report states," We need to get serious about moving away from burning fossil fuels, climate change is still the most serious threat to the nation's air quality and the health of people."

Rising carbon levels and higher temperatures are causing trees and plants to produce more pollen for longer periods during the year. As a results pollen amounts are as much as 40% higher than they were in the past decade. This leads to increased levels of discomfort and more severe allergies.

New York Times: The drying up of Utah's Great Salt Lake and the Salton Sea in Imperial Valley, California, if allowed to continue, will create a condition that could turn the air around the Salt Lake and the Imperial Valley into poisonous clouds of dust containing arsenic and other air born dangerous heavy metals and contaminants created by past mining and farming activity. Just the polluted dust blown in the air from these dried-up lake beds can cause serious respiratory problems.

Professor Benjamin Abbott an ecosystem professor of ecology at Bingham Young University recently declared that the Great Salt Lake could disappear in five years unless there is a dramatic increase of water flow into the lake. Decades of overconsumption of water, a major drought made worse by climate change, threatens to shrink the lake and cause harm to the region's public health and economy, Abbott stated. The lake has hit new record lows for the second time in less than a year. In 2021, the lake reached its lowest level in history at 941 square miles down from a peak of about 2,400 square miles in 1986-87 according to the geological survey. This is 19 feet below the average level the lake has been since 1850. Today its salinity is five times saltier than the ocean.

It will become a tragedy if further drying occurs. Roughly 350 bird species depend on its ecosystem's; it provides $2.5 billion dollars in direct economic production from mineral extraction, recreation and brine shrimp harvesting that employs 9000 people. The waters suppress heavy metals and cyanotoxins that have accumulated in sediments over hundreds of years. When sediments are exposed dust from the dry lake bed can blow them all over the country. This dust has already been observed as far away as Wyoming and Arizona.

Salt Lake City and its suburbs is one of the fastest urban growth areas in the country and its water supplies are starting to dwindle. Far less water now makes its way down from the adjoining mountains into the Salt Lake due to rising temperatures and less snowfall. Water is becoming a progressively more expensive commodity in the Salt Lake valley with local officials becoming increasing desperate by starting to impose strict water conservation policies.

In California's Imperial Valley, the cities of El Centro, Imperial, Calexico, and other small farming communities will be exposed to windblown clouds of dangerous dust containing toxic carcinogenic pollution from a dried-up Salton Sea. This dust will also affect the millions of people living just below the border in Mexico. The sea that was at one time the largest inland body of water in California has been drying up and now has lost a third of its size in just 25 years and one day will revert to its historical condition of a dry lake bed. With summer temperatures of continuous 100-degree days creating high levels of evaporation and less fresh water flowing into the sea from agriculture drainage, that has over the years become increasingly saline, its water is approaching the salt content of the ocean. Once one of the finest warm water fisheries in the state it is no longer capable of supporting a fish population of any species. It is fast becoming a dead dry lake with serious consequences for the surrounding communities and the many bird species that depend on it.

Methane, Carbon Dioxide & Permafrost

One of the main problems the world has in regulating climate change is controlling the release of methane gas, a gas that is many times more damaging to the climate then carbon dioxide. Although this gas stays in the atmosphere for only about ten years as compared to carbon dioxide that stays around for hundreds of years its effect, because of its potency, can be severe. Across the Arctic there are areas with lakes where the water appears to boil. The cause of this natural phenomena is the frozen methane called hydrates that are being released from the bottom of the lake by the warming of the water and then appearing on the surface as bubbles as it is released into the atmosphere. If you were to light a match, it would burn.

Permafrost covers around nine million square miles of the Arctic. It has been determined that up to two thirds of the Arctic's near surface permafrost could be lost by 2100. The melting of the permafrost will influence every part of the planet. As it releases methane and CO2 it will also release nitrous oxide gas (N20), a powerful greenhouse gas that is almost 300 times more powerful than CO2. There are estimates that there are over sixty billion tons of all these gases locked in the permafrost that could be released rapidly from an area covering almost one fourth of the entire Arctic. This a small percentage of the almost 1,700 billion tons the area is estimated to contain that could be release continually over a long-time span. A continual loop could occur, as it is now with the with the Arctic Sea ice, where more thaws release more gas creating more heating that creates more thaws and releases.

In 2021 at the recent United Nations climate summit held in Glasgow Scotland (COP) two hundred of the world's nation met to discuss and act on the world's climate struggles. There were some positive results that came out of the conference. A majority of the attendees agreed to reduce methane emissions by 30% although some of the biggest emitters such as Russia, China and Australia refused to sign the agreement. There was a positive action and agreement to reduce forest destruction and renew and protect existing forest reserves. Unfortunately, Brazil still continues to allow and encourage huge areas of the Amazon Forest to be destroyed each year for timber resources, agriculture and cattle production. A proposal to phase out the use of coal, the major source of atmospheric warming was scuttled and filled with loopholes that effectively killed the proposal.

Although some positive results were obtained by the attendees, what all of these proposals suffer from is the inability to address the fact that if the world is to avoid the worst of the climate change tragedy, we cannot keep extracting and using fossil fuels. As long as national governments continue to listen to the interest of fossil fuel lobbyists, the efforts of the climate summits will continue to be riddled with loopholes that will delay and derail the ability to achieve meaningful reduction results. There were five big loopholes that were glaringly apparent as the conference ended.

1. **Subsidies and Finance:**

   Governments continue to subsidize the fossil fuel industry with fossil fuel subsidies globally running at 11 million dollars every minute. These government subsidies make it difficult to halt emissions because subsidizing the cost of production and sale of fossil fuels continue to make the industry feasible. As an example, banks have recently provided $575 billion in fossil fuel finance to some of the world's largest polluters. Countries such as Australia should terminate new subsides such as the National Party's proposal for a new rail line to the Gladstone coal fields.

2. **New Production:**

   Despite overwhelming evidence that most of the world's remaining fossil fuels reserves should remain in the ground, governments are still approving new projects. The United Kingdom, despite being the host of the Glasgow conference, has 40 fossil fuel projects in the pipeline. Australia also continues approve new coal and gas developments. The government has approved eight new projects since 2018. Until climate negotiations put a ban on new fossil fuel projects and agree to a clear and rapid phase out of current production levels, the fossil fuel industry will continue to thrive and the earth will continue to heat.

3. Business as usual:

Another loophole for the fossil fuel industry is how it is being allowed to continued its huge levels of production because it has committed (in some cases) to making its operation "green." Measures such as carbon capture, tree planting, storage and offsetting have been touted by some governments and oil corporations as solutions to reducing the industries emissions. But these are not real solutions, they simply allow fossil fuel production to continue. These measures are not a substitute for genuine reductions to fossil fuel use and misleadingly give the impression that the industry is reforming and reducing emissions. It's a fraud being practiced against the world.

4. Influence:

The loopholes that allow and encourage fossil fuel production are no accident. The largest group of representatives attending the conference were lobbyists from the fossil fuel industry. The companies have exerted tremendous pressure on governments decision making process in their approach to climate change. As an example, the industry's influence on successive Australian governments has been well documented with over $136.8 million in political donations between 1999 and 2019.

5. **Decoupling production:**
    The failure to address these loopholes will mean that the production of fossil fuels will continue for much longer than it should. The fact that there are still willing buyers for fossil fuel assets such as the Australian coal mines indicates investors are anticipating years of profits and few climate liabilities, from fossil fuels, despite the efforts taken at the Glasgow conference. A key problem is ending the false idea that progress can be achieved in cutting emissions while simultaneously supporting fossil development. If countries can't agree on cutting support for production, then the loopholes will continue that will allow the industry to flourish.

And now we have another climate conference with 193 attendees in 2022. Will the results of these meetings be any different than in 2021? The world is watching.

A methane feedback loop that is beyond humans' ability to control may have already started according to the National Oceanic and Atmosphere Administration (NOAA). According to NOAA methane is 25 times more powerful at trapping heat in the atmosphere as compared to carbon dioxide. It has a huge effect on increasing the rate of climate change.

Xin Lan, a research scientist at NOAA, in an interview with Newsweek, disclosed that the majority of methane emissions after 2006, were caused by natural wetlands and man-made emissions. Natural wetlands, tundra and permafrost produce methane when organic matter decays caused by increasing rainfall, thawing and climate change heating; man-made emissions are caused by livestock, waste, algae, and landfills. Lan said "because of Earth's climate is already warming the methane produced by permafrost and wetlands is set to increase because of thawing. This signals the beginning of a feedback loop, an ongoing cycle of heating and thawing that cannot be broken." A continuing loop could produce an endless amount of this gas.

Scientist have discovered a vast methane leak at a Pemex oil field in Mexico. Using satellites, the leak was determined to have occurred at an offshore oil drilling platform in the Gulf of Mexico during six days from Aug. 5th. to Aug. 29th. During those days of leaking a total of 44,064 tons of methane were released into the atmosphere. This is the equivalent of 3.7 tons of CO_2 and was the second major leak from the same oil field following a major leak in Dec.2021.

One of the largest methane sources is permafrost that is becoming less stable as the weather warms in the Arctic. Permafrost is ground that has been permanently frozen for hundreds and thousands of years. Locked within this frozen ground is organic matter, the remains of plants and animals from an earlier period in earths long history when the area was much warmer. When it thaws, as it has already started in many areas of the Arctic, the organic matter defrosts and starts to break down, releasing its carbon.

Research published by the National Oceanic and Atmospheric Administration (NOAA) shows that 2021 had the largest annual increase in atmospheric methane since measurements were began in 1983. NOAA said that the annual increase in atmospheric methane during 2021 was 17 parts per billion, the largest amount of yearly increase since measurements were started. This should not be a surprise since the Arctic temperatures have been increasing at the highest rate in the world for several years. A report by the World Weather Attribution said that by 2050, temperatures in Siberia may increase by 2.5 degrees Celsius. Much of the infrastructure in the Arctic areas has been built on the permafrost that for years has been stable and considered permanently frozen; it is stable no longer.

Levels of carbon dioxide (CO2) also increased at historically high rates in 2021, NOAA reported the global surface average for carbon dioxide during 2021 was 415 parts per million (ppm), an increase of 2.66 ppm over the 2020 average. This was the 10th consecutive year CO2 increase of more than 2ppm. "Our data show that global emissions continue to move in the wrong direction at a rapid pace," NOAA Administrator Rick Spinrad said in a statement. "The evidence is consistent, alarming and undeniable."

The burning of fossil fuels such as coal, oil and gas releases greenhouse gases which has caused the temperature of Earth's atmosphere to rise to levels that cannot be explained by natural causes, scientist say. Carbon dioxide traps solar radiation and maintains it in the atmosphere. It is invisible, order less and colorless yet it is responsible for 63% of the warming for all greenhouse gases, according to NOAA's Earth Systems Research Laboratory in Colorado.

Today's atmospheric levels of CO2 are similar to where they were during the mid- Pliocene period, about 4.3 million years ago according to NOAA. During that period sea levels were approximately 75 feet higher than they are today and the average temperature was 7 degrees Fahrenheit higher. Large areas of forest covered the Arctic that today are now frozen tundra.

About 36 billion tons of CO2 were emitted into the atmosphere in 2021 by human activity; roughly 640 million tons of methane were emitted during the same period. Methane stays in the atmosphere for about ten years; some of the CO2 emitted today will warm the planet for hundreds of years and it will continue to accumulate as we continue burning.

A study by scientist that was reported in the Journal Nature found that the thawing permafrost will start destroying the infrastructure of buildings, highways, pipelines and other structures by 2050. But current evidence has shown this effect has already started in some areas. Recently a fuel spill from a power plant in the Krasnoyarsk region of Siberia was caused by the thawing permafrost and there are many other examples of structure failures due to defrosting. It is believed that the ground beneath the Nor-nickel thermoelectric plant became unstable causing the accident.

The study shows "That nearly four million people and 70% of the current Arctic infrastructure occurs in areas with high potential for future thawing of the near surface permafrost," as reported in the journal Nature. What is alarming is that these figures are not reduced substantially even if the climate change target reductions are reached. Reuters: Russian Central Arctic. The thaw of permafrost is crumbling buildings, threatening more than a fifth of Russian infrastructure and a critical part of its economy. Two thirds of the country sit on soil once thought to be permanently frozen, including most of its oil and gas infrastructure. In the town of Yakutsk residences describe water pipes that regularly burst and holes occurring in buildings and roads bucklering. The melting of Russia's long frozen tundra could frustrate efforts to curb climate warming emissions. In the near past it was believed that the melting permafrost would not have an impact until the end of the century, now that has been proven wrong.

Many mines, commercial buildings and industrial plants are experiencing increasing corrosive leaks and cracks. Pipeline support systems twist and bend when the supporting earth settles. The Russian government says that 40% of its buildings and infrastructure in permafrost -covered areas have already been damaged. "There isn't a single settlement in the Russia's Arctic where you wouldn't find a destroyed or damaged building," said Alexey Maslakav, a scientist at the Moscow State University. Pipelines, storage facilities, electric power systems and roads are increasingly in need of repair. As reported by the Wall Street Journal 10/5/2021.

Permafrost could also release long buried pathogens that the world has never experienced potentially harming wildlife and humans. "It is important to understand the secondary impacts of these large-scale earth changes such as permafrost thaw," said Kimberly Minor, a climate scientist at NASA's Jet Propulsion Laboratory. There is good reason to be wary of pathogens emitted from the ice. Recently a team of scientist discovered 28 new novel viruses in a melting glacier in Tibet, and 100 pathogens have been found in the deep permafrost in Siberia.

The Arctic permafrost area is estimated to store around 1,700 billion metric tons of carbon including carbon dioxide and methane. This is more than twice the amount of carbon currently in the atmosphere. If this amount was ever completely released it would be equal to more than 50 times the amount of fossil fuel emissions the world produced in 2019 according to a report by NASA.

Kimberley Miner, a climate researcher at NASA's Jet Propulsion Laboratory in Southern California is studying microbes (germs) frozen in the permafrost for thousands of years. Human contact with these biological hazards in the future is assured since many industrial sites, homes, apartments and military projects have all been built on permafrost in recent decades. Miner said, "that scientist have little evidence of what may occur when they are released into our ecosystems. We have a very small understanding of the kind of microbes that have existed in many different conditions frozen in time periods beyond our knowledge that may have the potential to re-emerge, and how they could affect mankind."

From another study published in the journal Nature Climate Change: By 2040 Northern Europe and Western Siberia will not be cold enough to sustain the permanent permafrost. These areas are fast approaching a tipping point that will release billions of tons of carbon. These frozen permafrost areas store twice the amount of carbon as all of the European forest combined; as the heating continues it will be released. It is vitally important that these ecosystems are understood and considered on their impact of climate change.

Methane Detection: An analysis of data from sensors aboard a Canadian satellite have for the first time identified an individual farm as the source of methane emissions from cattle. The aerospace firm GHGSat used one of its three orbiters for the discovery, which demonstrated a new level of precision in identifying the powerful greenhouse gas polluting the atmosphere. With high-resolution images from Feb. 2, 2022, the firm used wind modeling to trace the source of the methane from bovine flatulence and belching to a farm near Bakersfield, California. The new technology could help regulatory agencies monitor how much methane is being generated by specific cattle ranches and other sources.

Economics and Politics

The accumulation of all powers, legislative, executive and judiciary in the same hands whether of one, a few or many, self-appointed or elected may justly be pronounced the very definition of tyranny. James Madison

Scientists can advise, warn, and plead as they have been doing for many years and to frequently they have been ignored, maligned, and accused of being extremist and fearmongers. Until recently they have had very little acknowledgment for their efforts. They have had to contend with complete denials of the coming emergency and the supporting truths they have presented. Unfortunately, today many of their findings and projections are being realized the hard way, with increasingly severe climate events. But there are still those, many in powerful influential positions, that question the scientific facts and refuse, because of political and financial interest, to take action to limit the continued release of harmful pollutants.

The only power the scientist have is public opinion and even that, until very recently, has many times been ignored by lack of concern or knowledge. It is a shame that many people today are having to learn of climate change and the scientific truths the hard way by suffering the loss of life and property by damaging weather episodes. The scientists do not have the power of the purse, or the political leverage to overcome the directed special interest and millions of dollars targeted by the energy industry at the decision makers who are ignorant of what is occurring or are only interested in obtaining and holding power regardless of what is becoming more apparent as the world burns.

An international coalition of more than 1,400 scientists recently signed an initiative declaring that the world's leaders are consistently failing to cope with the main cause of climate change and the deepening climate emergency. It will be the politicians and the moneyed interest that will ultimately determine the fate of the world and billions of its people. That is a very scary statement and let's hope it's not true, but as of today it's as close to reality as you can get.

What we are considering is the slow destruction of life as we know it. Our habit of living beyond the laws of nature are no longer sustainable. Mother delivers the rules and she is having her way. Extinction? Not tomorrow or next year or the next hundreds or even a thousand years or more. She is in no hurry. There are so many things to go wrong that it takes time. Extinction is a gradual, slow, painful process involving many different destructive outcomes occurring over very long time periods, but once that process is accelerated through natural or human caused events such as climate change the clock starts ticking.

Recently the International Governmental Panel on Climate Change (IPCC) issued its latest report on efforts to control global warming. It showed a "listing of broken climate promises that was putting mankind on a track towards an unlivable world," in the words of U.N. Secretary-General Antonio Guterres.

The 2022 Climate Change Conference, COP27 held in Egypt. In his opening remarks at the COP27 United Nations climate conference in Egypt, Antonio Guterres said the world was on a "high-way to climate hell with our foot on the accelerator." He said in the past few weeks scientific reports have painted "a clear and bleak picture" of global warming with greenhouse gas emissions still growing at record levels instead of going down 45% by 2030 as scientist say must happen. He also told leaders to "co-operate or perish," singling out the two biggest polluting countries, China and the United States The delegates from 193 countries attending this conference have before them this most recent report from the world's scientific community. The report represents the scientific consensus on climate change, its causes and impacts.

A huge example of indifference: China is building dozens of coal fired power plants even as the ruling Communist Party promises that their emissions of climate- changing carbon will peak by 2030. China relies on coal for two-thirds of its power. It emits more carbon than the United States and India combined. China's president increased coal mining after power shortages last year caused blackouts and factory shutdowns. They added 7,500 megawatts of coal-fired power plants in the first half of 2022, four times as much as No. 2 India according to Global Energy Monitor. Another 300,000 megawatts are under construction or planed. China's action on climate change is" highly insufficient" says the nonprofit group Climate Analytics. An understatement if there ever was one.

Some of the significant findings of the report:
The physical basis for climate change is blamed on humans for the rising temperatures. Climate change is dangerously close to spinning out of control. For the first time it called for urgent action to control the release of methane gas.

The report stated that the world's nations, including the wealthiest, need to prepare for climate impacts and start adapting to a coming warmer world including the impacts of more frequent and severe heatwaves, stronger storms and higher sea levels. Millions of people will face poverty and food insecurity in the coming years as the increased warming affects crops, water supplies and the disruption to trade and labor markets.

"It's now or never, "one of the report's authors said in releasing findings that show that only drastic emission cuts now would prevent a future catastrophic uncontrollable climate. The energy transition to renewable sources and clean burning fuels is moving too slowly. It urged strong climate action in agriculture where farming methods and better forest protection could help curb emissions. It also warns that climate change will seriously threaten the world's economic growth.

All of these recommendations for action submitted to the 193 members of the climate conference come from the most reliable scientific information available to the world's leaders. How will they respond to this wake-up call for action? Will it be a meeting where decisive action is taken on climate change, or will it be a distressing example of international greenwashing that makes little serious progress as has happen in the past?

The report is an indictment of governments, corporations and investors who have signed on to the "net zero" carbon goal but have put little or no concrete effort behind their pledges. "Governments and business are saying one thing and doing another," Guterres said. "Simply put, they are lying." Scientists have concluded that the current world investments to mitigate global warming; "are three to six times to low," then what is needed even to hold global warming below the two degrees Celsius target by 2030. At this meeting the countries were to deliver reports and plans as to how they were to lower their greenhouse gas emissions as they pledged to do at last year's conference. To date, only 23 out of the 193 countries have submitted their plans to the U.N.

One of the world's largest oil producers, Saudi Arabia, announced that it aims to reach "net zero" greenhouse emissions by 2060.

Although the kingdom said it will attempt to reduce emissions within its own borders, there is no indication it will slow down investments in oil and gas or relinquish its sway over the worlds energy markets by moving away from the production of fossil fuels. Saudi Arabia says it will reach net-zero through a so-called "Carbon Circular Economy," approach, which advocates "reduce, reuse, recycle and remove." It is not a valid strategy according to climate activists and scientist because it relays on still unproven carbon capture and storage technologies that have not been developed, rather than reducing the use of fossil fuels.

While the new law recently passed by Congress was wildly heralded as the largest investment in climate change spending in history a recent poll shows that 49% of Americans say it won't make much difference on climate change, 33% say it will help and 14% think it will do more harm than good. As part of the poll 61% say they know nothing about the new climate effort.

The following are some projections and observations that paint a grim picture of progress:

Extreme weather is already causing enormous damage, and it will get worse. Since 1980 there have been 323-billion-dollar weather and climate disasters in the United States, coasting nearly $2.2 trillion. In 2021 the U.S. spent $145 billion on its worse 20 climate and weather disasters.

The Office of Management and Budget (OMB) has said unaddressed climate change could cost the U.S. $2 trillion annually this century, not counting substantial cost to business and public health. The insurance company Swiss Re estimates un- migrated climate change could cost the world economy as much as $23 trillion a year by 2050.

Recent data collected by the National Oceanic Atmospheric Administration (NOAA) reports the costs of droughts, extreme weather events, floods, crop failures and other climate events in the U.S. alone was 145 billion dollars in 2021. And now, after the recent destructive storm Ida, the 5th strongest hurricane in history, hit Florida and the Carolinas, the initial cost estimates of the damage approaches 100 billion dollars, which may be conservative. Over the last 5 year the costs of these events have totaled $764.9 billion dollars not counting the present year. This is more than a third of the combined costs over the previous 42 years. How much will these costs increase in dollars and lives in the coming years?

According to the Lyndall Center for Climate Change Research, a global temperature increases of 3.7 degrees Celsius by the end of the century would cause damages of $551 trillion. That is more than all the wealth on the planet today.

The First Street Foundation, a research group, reports that more than 700,000 U.S. apartment buildings, malls and office complexes are at risk this year of floods severe enough to impede access.

Business downtime could reach 3 million days and cost $50 billion a year. According to a poll taken in the summer of 2022 by the New York Times, a majority of Americans surveyed now believe that our political system is to divided to solve the nation's problems. We are awash in violence and social disorder. The nonprofit Gun Violence Archive has documented 429 mass shootings from January to September, so far this year. "Mass shootings" are defined where at least four people are killed or injured. Meanwhile the Supreme Court by overturning Roe v. Wade has led to stark divisions in the country. Our political gulf is a deep and shattered mess where Republicans and Democrats increasingly live in a separate media universe. Will todays petty politics destroy us?

"The alternate domination of one factor over another shaped by the spirit of revenge is itself a frightful despotism, sooner or later the chief of some prevailing faction, more able or more fortunate than his competitors turn this deposit to the purpose of his own elevation on the ruins of public liberty." George Washington's farewell address. Two hundred and twenty-four years ago Washington predicted Trump and his Republican cohorts.

What are some of the empty promises that the U.N. Secretary has criticized? Researchers at Japan's Kiyoto University studied the corporate reports and spending of British Petroleum, Shell, Chevron and Exxon Mobil. They concluded that "no major oil company is currently on the way to a clean energy transition."

Big investors are also part of the problem. Tim Quinson, senior executive editor at Bloomberg, reports; "None of the world's largest asset managers has called on the fossil-fuel companies to stop the development of new oil, gas and coal projects. Notably, 30 of the largest asset managers have invested $550 billion dollars in oil, gas and coal expansion.

Climate Action 100, a group of investors pushing 166 corporations to act faster against the effects of climate change, reports that only 17% have established emission-cutting targets. The list of empty promises continues and so does our march to an irreversible climate catastrophe. Guterres called the empty promises a "file of shame."

The United Nations General Assembly 2022 meeting opened with dire assessments of a planet beset by escalating crisis. The tone of the meeting was far from being encouraging, indicating a tense and worried world. "We are gridlocked in a colossal global dysfunction and the world is heading in the wrong direction," Secretary General Antonio Guterres said in his opening remarks to the assembly. Without considerably more effort to limit greenhouse gas emissions, the physical and social economic impacts will become increasingly more devastating. The report states that weather related disasters have increased five-fold over the last 50 years and are sure to worsen. The past seven years were the Earth's warmest on record and there is a 93% chance that at least one year in the next five years will again see record heat. Guterres said, "There is nothing natural about the size of these disasters. They are the price of humanities fossil fuel addiction. Yet each year we double down on fossil fuel obsession, even as the symptoms get worse."

Rich energy countries should be required to share some of the windfall profits to aid victims of climate change. The fossil fuel industry, which is responsible for a large share of the planets warming gases is "feasting on hundreds of billions of dollars of windfall profits while household budgets shrink and our planet burns," he said.

He urged richer countries to "tax profits of energy companies and redirect the funds to the damage caused by the climate crisis." Oil companies in July 2022 reported huge excessive profits of billions of dollars per month. Exxon Mobil posted three months profit of $17.85 billion; Chevron of$11.62 billion and Shell of $11.5 billion. "It is high time to put fossil fuel producers and investors on notice. Polluters must pay," Guterres said, censuring them for their massive public relations campaigns.

The U.S. Treasury Secretary, Janet Yellen warned in a recent speech of an economic calamity if climate change is not addressed with immediate government action. Yellen said the increasing frequency and severity of disasters caused by climate change could create devastating short-term supply reductions of every day goods and services that soon could become commonplace.

In a new Yahoo News /Gov. poll that asked Democrats to name the most important issue to them when thinking about this year's 2022 mid-term elections, only 5% said climate change. In a similar 2021 poll report by Yahoo, 67% of Republicans said climate change "is not an emergency." Yet the consequences of climate warming can be seen occurring everywhere in the world. Millions of people are suffering today from its effects and yet we as a nation are unable to come together for the simplest of actions by stopping the release of damaging soot and warming gases into our environment from coal fired power plants. The atmospheric increase of concentrations of carbon dioxide and methane continue to set new records and it won't slow down until enough members of Congress are elected who are willing to reinstate and enforce the regulations just suspended by the Supreme Court.

The recent U.S. Supreme Court decision has severely handcuffed the country's regulatory ability to reduce the countries greenhouse gas emissions. Unfortunately, the landmark legislation recently passed by Congress and signed by the President will fall far short of meeting the countries emission reductions largely because of the high court's action. This raises some profound questions about the resolve and our ability to overcome the influence of the capital markets, the courts and the corporate money influences if we are to forcefully address the climate crisis.

Yahoo News, The Supreme Court's Ruling

In a majority opinion written by Chief Justice John Roberts, the Supreme Court conceded that the Obama EPA standard called Clean Power Plan was intended to curb greenhouse gas pollution that is causing climate change and which scientist have long warned threatens life on earth as we know it.

"Capping carbon dioxide emissions at a level that will force a nationwide transition away from the use of coal to generate electricity may be a sensible solution to the crisis of the day," John Roberts the Supreme Courts Chief Justice wrote. Then the big "but" asserted itself and after a moment of lost common sense he wrote: "But it is not plausible that Congress gave the Environmental Protection Agency authority to adopt on its own such a regulation scheme," he concluded. The court sided with the state of West Virginia, a large coal producing state, saying the EPA had overstepped its authority in trying to enforce the clean air act by limiting coal fired power plant emissions.

Does anyone else see the hypocrisy in this political ruling?

As Justice Elena Kagan put it in her dissent of the ruling, the court; "appoints itself instead of Congress or the expert agency, the decision marker on climate policy. I cannot think of many things more frighting."

"There can be no liberty if the power of judging is not clearly and completely separated from the legislative and executive functions. Having neither the power of the purse or the sward, a free people cannot and will not exist without the complete independence of the judiciary." Alexander Hamilton

"This court is deliberately handcuffing our ability to respond to the most serious crisis humankind has ever faced," Lisa Graves, executive director of True North Research and a former Department of Justice official said in a written statement. "And it is doing so recklessly during the most crucial decade we have to limit carbons impact on the heating of our planet, the sustainability of our oceans and agriculture, and our ability as a people to survive the climate changes that today are occurring." This decision will seriously hamper the country's ability to reduce its emissions target for containing warming to 1.5 degrees Celsius by 2030.

Naomi Youder, a scientist with Healthy Gulf, a New Orleans based environmental group said she has "close to zero trust that the private sector will adjust their business models to significantly address climate change. Though some company's say they take the problem seriously, they continue to invest in fossil fuel development."

Many experts argue that even if companies are sincere about changing to adjust to changing climate there is no way to enforce the changes without government involvement to insure the necessary emissions reductions. At a meeting of the leading U.S. banks this year, a shareholder resolution was presented that would have committed the banks to making no new investments in the development of fossil fuels. The resolution failed, gaining only 11% of the votes. Blackrock the huge investment firm, fearing resistance from conservative states with large pension funds containing energy investments, such as Texas, gave its support to the oil and gas industry and resisted calls for no new investments.

In a report prepared in 2021 the International Energy Association (IEA), an intergovernmental agency based in Paris, said there should be no new coal, oil or gas field developments in order to meet the climate targets limiting global warming to 2.7 degrees Fahrenheit. The number of countries pledging to reduce emissions in the coming decades has grown the IEA report said. But the pledges to date, even if fully achieved fall well short of what is required. Governments must moderate the system by creating incentives or penalties that account for climate damage.

Many economists call for a tax on carbon to try to encourage companies to reduce their emission by putting a price on polluting the atmosphere. Canada, China, Japan the European Union and the United Kingdom have some form of a carbon tax. We, the largest contributor to the heating of the planet has none. Congress voted against such a tax under extreme pressure from the energy companies in 2010, and has repletely resisted, by political pressure, for such a move. With a conservative congress and the energy companies having bought them, what is the chance of this ever occurring?

The Securities Exchange Commission (SEC) has proposed a rule to require companies to disclose their carbon emissions and climate related business risks. If, and that's a big if, adopted it would expose polluting companies to public outrage or scare off investors.

Investment decisions using environmental, social and government considerations have been applied for decades. But recently such practices have been described as "Woke Capitalism" (ESG) by powerful politicians, and a concerted effort has been waged to stop companies and government entities from taking these factors into consideration when making investment decisions. "Woke," has become a political battle cry. Eighteen states have either proposed or adopted rules that limits the ability of the state government and public retirement plans to do business with firms they believe do business with environmental, social and government entities. "It's a sinister lie that's deeply counterproductive, not just to the climate but also to people's pocketbooks and pensions," said Alicia Seiger, who teaches energy finance at Stanford University's law and graduate school of business in California.

ESG investing is seen as a threat to the oil, gas and coal industries as the Republicans in red states fight the transition from fossil fuels. They see those companies that practice ESG often making efforts to reduce their carbon footprints. This has become appealing to investors who are making decisions to invest. A group of 19 state attorney generals are investigating the role of banks who invest in companies that are reducing greenhouse gas emissions. They claim the banks are favoring companies that follow a "Woke" climate agenda. Republicans are also passing laws in a number of states that prohibit state investments in companies that practice ESG. Three of the largest financial companies in the country that have invested in ESG for many years, BlackRock, Vanguard and State Street issued the following statement: "Climate change, and the ongoing global response, will have far-reaching economic consequences for companies, financial markets, and investors, that present a clear financial risk." A survey from Penn States Center found that 63% of voters surveyed said government should not set limits on ESG investments. "Our research found that neither Republican or Democrat voters support policies or legislative efforts to curb ESG investing. The consensus among voters surveyed was that companies should be able to decide to invest in ESG that benefit society without government interference."

Millions of Generation X and other young investors around the world have contributed to ESG investing. These investments try to improve society and hold businesses to high standards; they have become an influential financial force. Over the past decades, ESG has grown into a trillion-dollar enterprise that has had real-world impact on the economy, society and specific issues investors care about like climate change. Socially responsible investments are expected to reach $50 trillion by 2025 up from $30.7 trillion in 2018. One prominent opponent of ESG is Florida's Republican Governor Ron DeSantis. He chaired a state board that banned investment fund managers from considering social, political or ideological interest (climate change) when making decisions for Florida's retirement system, calling it "woke investing." Texas and West Virginia are also taking a similar approach. Texas has banned state and local government agencies from doing business with financial firms that refuse to support the gun and fossil fuel industries. West Virginia is focused on financial firms that oppose coal investing.

Cities and states have the power to control climate policies. This year, Pennsylvania became the latest state to enter the Regional Greenhouse Gas Initiative, joining 11 other Northeast states in an agreement that caps carbon emissions from power plants. More than 100 cities have committed to achieve net zero emissions in the coming decades, and Steven Schooner, a law professor at George Washington University, notes that cities and states have more purchasing power than the federal government. He said that enables the cities and states to help drive a renewable energy transition by providing a dependable market for sustainable supplies.

A recent Bank of America Global Research Report found that the consequences of climate change will be dire and extreme. "This is the last decade to act," the authors wrote. Absolute water scarcity is likely for 1.8 billion people, 100 million face poverty and 800 million are at risk of rising sea levels by 2025 according to the report. Migration caused by climate events could move 143 million from poorer countries driven by warmer weather and scarce water.

The cost to reduce carbon releases to sustainable levels could reach 5 trillion dollars a year the reports states. It is the annual amount that will be the average for the next 30 years. A massive number of 150 trillion dollars, said Hein Israel, Director of research at Bank of America/Merrill Lynch. That number is almost twice the global GDP in 2019. This cost is for transition of emitting industries into cleaner industries.

Who in their right mind thinks this is going to be possible without destroying the world's economy? But if the release of additional vast amounts of carbon into the atmosphere continue, these costs would appear to be a pittance in comparison to the damage and hurt the world would suffer. Again, the dilemma, we can't afford these tremendous transition costs, yet we cannot afford not to afford them; so where does that leave the planet and its people?

Regardless of the report's projections, the high costs for the climate disasters and drought are already occurring and within a few decades the total costs will be astronomical as warming continues, migrations increase, and industrial upheaval accelerates. The warming climate will affect every existing service from water and sewage treatment, mass transit, food distribution, and health care while eroding the wealth of millions of people.

Texas, Austin: The state's largest oil lobby is mounting a grassroots effort designed to pressure the Environmental Protection Agency (EPA) into not implementing new air pollution regulations.

Climate Insurance: The floods, wildfires and other weather- related disasters that have recently plagued parts of southwest Australia may soon become so frequent and vast that many homes will be deemed "uninsurable." This means that the risk is so great that insurance is only available at such high cost that no one can afford it.

In America, climate change is increasingly impacting working class and minority communities. A growing number of American families are facing financial ruin as the available private insurance and government relief programs fail to keep up with the rate of weather-related events intensified by climate changes such as the most recent storms.

Democracy destroys itself because it abuses its right to freedom and equality. Because it teaches its citizens to consider audacity as a right, lawlessness as freedom, abrasive speech as equality and anarchy as progress. Isocrates 436-338 BC

Rivers, Water and Flooding

Parts of Sidney Australia largest metropolitan area has just suffered its fourth major flood emergency since March 2022, that forced thousands of flood-weary residents to leave their homes again. Dams have overflowed because of continuing intensive rainfall. Sidney neighborhoods have had torrents of water flooding their homes. Residents are tired of the months of constant storms and the threat to their homes and lives because of the changing weather patterns caused by the shifting climate.

Long before Climate change was a concerned, California experienced The Great Flood of 1862 that stretched up to 300 miles long and 60 miles across. A recent study found that a similar flood today would displace 5 million to 10 million people, cut off water and power supplies, disrupt and destroy the state's major freeways for weeks or months and cause massive economic damage and submerge major Central Valley cities as well as large parts of the Los Angeles basin. Could this winter's state wide intense storms with their extreme rainfall and widespread flooding be a forerunner of a larger future event?

Mega droughts are today, the areas main concern along with wildfires and earthquakes but the study indicates that a new crisis is looming in the state: mega floods. Because of the warming climate there is considerably more moisture circulating in the atmosphere, increasing the likelihood of a major storm system that could last several weeks and bring more than 100 inches of rain to hundreds of square miles of coastal and inland valleys. Similar major storms have happened in the past before the region became home to tens of millions of people.

With climate change each degree of warming dramatically increases the odds and size of the next mega flood according to the study. There is no such thing anymore of a 500-year storm or even a 100-year storm. The timing for these destructive events to happen has been greatly reduced. When floods occur in a warmer planet, the storm sequence is larger in almost every respect, said Daniel Swain, a UCLA climate scientist and co-author of the study. "there's more rain overall, more intense rain on an hourly basis, the storms move slower; and stronger winds." We are seeing that today in Pakistan where one third of the country is flooded and millions of people have been displaced. Swain said that such statewide storms have occurred every century or two in California over the past millennia and today the possibility of such an event has been amplified by the warming and saturated atmosphere.

It is estimated that such a flood that happened in 1862 would be a trillion-dollar disaster today. Parts of cities such as Sacramento, Stockton, Fresno and Los Angeles would be under water. It would be one of the most destructive catastrophes that has ever occurred in the world's history.

The realities of climate change were on full display the summer of 2021. At least 136 daily rainfall records were set during storms across five states along the Mississippi River. Detroit Michigan was swamped by one storm that drooped 7 inches of rain. Tropical storms soaked an area of the South flooding homes in Louisiana and Alabama where it drooped 8 inches of rain and claimed 14 lives. In August 2022, a Dallas County Texas rain gage recorded more than 14.9 inches of rain; several other rain gages reported 10 inches or more. A record breaking 3.01 inches of rain fell in one hour at the Dallas-Fort Worth International Airport. The warming atmosphere is capable of holding more moisture and these extreme flooding events are expected to increase. While in the west drought combined with soaring temperatures shattered century old heat records and prompted heat alerts, that killed more than 200 people and intensive wildfires spread over several states.

This year 2022, the Mississippi river in many areas has receded to the lowest level it has been in a decade. This river provides drinking water to around 20 million people. It also is the primary mode of transportation, carrying about 500 million tons of cargo each year. Its river basin is home to 57% of the farmland that produces 60 % of the U.S. grains and 54% of its soybeans. The river flows through ten states and fortunately not all of them have been affected by the drought and low flows. Louisiana has been the hit hard as it sets at the mouth of the Mississippi River Delta. It has been impacted by salt water intrusion and some communities have issued water advisories and may have to plan for desalination treatments to insure a reliable drinking water supply in the future. At one point on the river there were 2000 barges stalled due to the low water levels this year.

The increased heat has changed how moisture moves across the country. Scientist say it alters the flow of the jet stream, extends droughts and increases evaporation from the land and oceans. East of the Rocky Mountains more rain is falling with more intense storms. In the West the opposite has been true. People are experiencing fewer storms with long time periods between them.

Snow and water availability: In the future, snow free seasons are expected to last longer putting water supplies at risk. One of the most precious things for human survival may soon face challenges as climate change continues to modify the amount of snow that upon melting becomes drinking water. The warming is expected to disturb the state of snow-dominated ecosystems but the precise impact is at this time unclear according to a study published in the Proceedings of the National Academy of Scientist.

As the higher elevations warm, more moisture in the form of rain will fall reducing the amount of snow that is available for slow-melting and long-term runoff and storage. It will reduce the available water as a stored resource in reservoirs that furnish water for domestic use, power generation and agriculture. In the past, snow has been one of the best predictors of water availability in places like the Western U.S., where about two thirds of the water in streams and rivers comes from snowmelt.

Snowpack reductions, early snowmelt and runoff, lengthy growing seasons, increasing dry soils and increased fire risk are expected under the intensifying high carbon emissions that will result in longer, hotter and drier summers and longer snow free seasons. It is predicted that an additional 82 snow- free days per year will occur in the Rocky Mountains by the end of the century.

Earlier Melt: Winter ice on more than 117 million lakes at higher latitudes is melting about eight days earlier due to global heating. Iestyn Woolway of Bangor University in Whales, said earlier melt is affecting plants and wildlife while distressing the local climate. Ice currently forms on more than half of the world's lakes, with 90% of them located north of 30 degrees latitude. Woodway says the lakes will eventually be free of ice 15 to 45 days earlier, which will warm the areas surrounding the lake between 2 and 6 degrees. Another recent study that analyzed more than sixty lakes in the Northern Hemisphere found that their water temperatures have steadily increased during the past 100 to 200 years. Over the last 25 years the water temperatures increases have exceeded any other period.

The impacts on humans and the ecosystems of many species will be widespread with peak snowpack in the Northern Hemisphere predicted to decline by over 50% by the end of the century according to the report. Cold water fisheries will be severely impacted. The Colorado River, that depends mostly on the melting snowpack in the Rockies, supplies the majority of the water to Lake Mead and Lake Powell, the two largest water storage facilities in the U.S. The river has lost 20% of its flow capacity over the last two decades. Those reservoirs supply water and power to forty million inhabitants in the Southwest. Glacier melting and runoff are also the primary source of water for hundreds of millions of people throughout the planet. Although their melting is measured on a longer time scale than snow, their continued melting and disappearance will have a devastating effect on a huge number of people in the future.

A new study recently published by the U.S. Geological Survey, Montana Fish and Wildlife, The University of Montana and the Journal of Science Advancement considered how climate change has and will affect rivers and streams on the state's fishable waterways. "Trout fisheries have an enormous cultural and economic importance in Montana and all of the western states." The sport could be vulnerable to future warming. The researchers found that climate extremes such as less snowfall, early melting runoff resulting in lower and warmer waters are shifting the abundance and distribution of trout. By 2080, the study predicted that 35% of the miles of trout stream habitat may be lost.

Angling pressure on the states trout fishable waters has doubled from 1983 to 2017. The non-resident angular fishing days has increased 280%. The Montana angling industry is now among one the most important parts of the state's tourism economy. The amount of tourist dollars spent annually in the state is valued at more than 750 million dollars. This reduction of available streams and rivers fishing miles in the future will increase angling concentrations and pressure on the available fisheries. It is projected that tourist fishing revenue will decline by 65% by the year 2040 and 76 % by 2080.

Nineteen states in the East doubled their previous number of extreme precipitation days from three a year to six. As these storms grow in frequency and severity the amount of annual rainfall has increased for more than half the nation. At some point over the last three years 27 states, all east of the Rocky Mountains have experienced their highest 30-year average rainfalls since record keeping began in 1895, according to NOAA data. A dozen states, including Ohio and Rhode Island saw five of their 10 wettest years in history over the past two decades. During the same time period, eight states, five in the West, had at least three record - dry years. Heat increases evaporation and draws more water into the air where if forms weather systems that can produce more intense storms. "For every 1.8 degrees of Fahrenheit warming, 7% more moisture is absorbed into the atmosphere," said David Easterling, director of NOAA's National Climate Assessment Technical Support Unit. This is one of the main causes of the storms increasing severity. The year 2022 was the 7th hurricane season in a row to produce a deadly landfall hurricane on the U.S. mainland. No other periods have as many consecutive lethal storms.

This trend of higher and more frequent extreme weather conditions started in the late 20th century as the amount of greenhouse gas increased in the atmosphere. We as a nation and the world must take immediate steps to reduce carbon emissions that increase weather extremes. Rob Moore with the National Resources Defense Council said; "given the increasing frequency of weather disasters the nation should be galvanized into taking action. And yet we are not incorporating what we know about the future into actions that could relieve the effects of droughts and flooding."

The Washington Post: According to a new analysis just published by the nonprofit research firm Climate Central, roughly 4.4 million acres of land and 650,000 individual properties in the U.S. will be below sea level by the year 2050, based on current emission levels. The land affected could increase to 9.1 million acres by the end of the century. That's a staggering amount of lost land, infrastructure destruction and costs. One island off the coast of Louisiana is already 98 percent under water. The impact of sea level rise is today obvious as some communities face the prospect of abandonment and struggle today with sunny day flooding due to the higher tides. Municipalities and individuals will be forced to confront huge costs for removing inundated structures and flooded septic systems and relocating entire communities and their infrastructure.

Mother Nature is always going to win, she has a bone to pick with the human race and we can't really blame her. We are not going to engineer our way out of this and spending billions of dollars for beach repair, sea walls barriers and other efforts will only delay the inevitable. Costal infrastructure will have to be relocated eventually and millions of people will have to move, there is really no other alternative. Parts of the Barrier Islands have retreated more than 200 feet in the last two decades and some beaches are losing about 13 feet a year according to the National Park Service. By 2100, 13 million Americans could be displaced and more than one trillion dollars of property permanently flooded. Again, we are using the arbitrary time, the end of the century, but Mother could care less about time thresholds. She will continue to change, to adjust and bring her world back into balance. Those of you who are reading this will not experience these events, but your children and their children will.

It should always be remembered that the forces of nature can only be held at bay, never vanquished.

As the sea continues to rise, daily high tides are pushed further inland, flooding low lying areas. According to the study, Louisiana is the most vulnerable state with the potential loss of 25,000 properties that could be completely flooded by 2050 with the loss of 2.5 million acres of land. People are not going to pay taxes on properties that are under water and will have no alternative but to abandoned them. Governments could be responsible for thousands of properties that get abandoned. In England the rising seas have already left some places, such as Fairbourne in Whales below the spring tide levels.

Florida, Fort Myers: The National Oceanic and Atmosphere Administration reports that by the turn of the century there will only be one day in the year when high tides flooding is not an issue in the Fort Myers area. The city will be flooded by incoming tides every day according to the latest tide predictions. How long will people be able to contend with these conditions? The tragic devastation and loss of life that occurred in September 2022, in the Fort Myers area, as a result of the hurricane Ida, is a clear example of with the future holds. The big question may be; will Fort Meyers survive to the end of the century, or should it?

But even before that happens, communities today are already dealing with having to repair street and road damage as well as outdated water and sewer systems. "How cities and county's respond to these current and coming risks is material to their future and their ability to repay debt and to protect credit ratings," the analysis stated. What remains undermined is how are communities across the country preparing for changes they know are coming and what this country and the rest of the world does to slow down the continuing heating of the planet.

Perce, Quebec Canada: Over the past ten years the citizens of Quebec's costal community Perce, a town of several thousand, have removed roads, houses and other infrastructure from the water's edge and relocated them further inland rather than to try to continue to protect themselves from the effect of erosion due the rising sea levels. As many as 500 structures have been relocated. They have even removed the protecting seawall and rock riprap that had been installed for many years. They install a new boardwalk father back from the water without the concrete protecting wall that had previously increased the waves fury. The community learned that the sea prevails when you try to wall it off. The idea was to retreat and move with the sea not fight against it, knowing they were not going to win. It was found that seawalls and stone riprap, rather than absorbing the waves energy created a backwash that collided with the incoming waves, creating a turbulence that chews away the shoreline, increasing erosion. In the Perce area where rigid protection had been in place over many years, the width of the beach had decreased by about 70%. A novel approach that many coastal communities will be faced with in coming years.

Pakistan Flooding, Summer 2022: One third of the country has been covered with water caused by massive monsoon rains and glacier melt. The floods have affected over 33 million people and destroyed or damaged 2 million homes. About half a million flood survivors are homeless, living in tents and makeshift structures. Thousands of doctors and medics have been deployed to battle the outbreak of disease such as typhoid, malaria and dengue fever. The cost of the destruction is estimated at 10 billion dollars, damaged to roads and farmland has been extensive.

Pakistan is the home to over 7200 glaciers; this is largest number of any country outside the polar regions. Rising temperatures linked to climate change, are melting many of them faster than previously projected, adding waters to the rivers and streams that are already flooding due the increased rainfall. Recent studies show that the snow and ice now being lost from the Himalayan glaciers is at a rate that is at least ten times higher than the average rate over the past century. Hundreds of millions of people depend on these waters for drinking and agriculture. Glaciers have been retreating worldwide since the 1990s; this is unprecedented in at least the last two millennia and gives a clear signal of the impacts of global heating. They are declining constantly at today's temperatures indicating they will not be safe with any additional warming. In the cold confines of Iceland 90% of their glaciers have continued to lose substantial amounts of their ice.

Drought

From an article appearing in the High-Country News,
by Theo Whitcomb

In late April, Oregon's Gov. Kate Brown increased the drought emergency status to cover more than half her state. Further south the Metropolitan Water District of Southern California which supplies water to millions of people, restricted outdoor water uses for the first time in its history. In Colorado, the U.S. Department of Agriculture designated the entire state a "primary natural disaster area" due to the threat of drought, this was also considered an exceptional action. Yes, the drought is real and becoming verry serious. All of the Southwest has been affected by dry conditions: Utah and New Mexico both issued separate emergency declarations. Shrinking snowpacks and rainfall have depleted reservoirs that show symptoms of the West's worst number of dry years since 800 A.D. There is significant likelihood that the existing megadrought will continue. A study published in Nature Climate Change in February predicted a 94% chance the drought will continue through 2023; the chance of it continuing through 2030 are 75% when considering the increasing impacts of a warming climate. According to the U.S. Drought Monitor, most of the West in in "moderate to severe drought." Certain regions, like eastern and southwestern Oregon, California's Central Valley, southern Nevada and eastern New Mexico are in "extreme" to "exceptional "drought.

The Southwest

Lake Powell and lake Mead, the two largest reservoirs in the nation, are at record lows. Lake Powell is just at 24 % of its capacity while Lake Mead is only a quarter full. Powell's stored supplies have fallen to just about 5-million-acre feet. The lakes full pool capacity is 26 million acre-feet. Cities from San Diego to Las Vegas are adapting emergency programs like water recycling and restricted outdoor watering.

The Federal Government will impose deep cuts in water allocations to Arizona and Nevada in 2023 as the reservoirs continue to decline. Even with substantial reductions this will leave the Colorado River and the 40 million people that depend on it an uncertain future. "Currently, the Colorado River does not have enough capacity, given the drought and climate change, for the present-day level of use," said John Entsminger of the Nevada Water Authority. "The magnitude of the problem is so large that every single water user must contribute solutions to this problem," he said.

In the Southwest 98% of the area is in drought, according to the U.S. Drought Monitor. NASA's Earth Observatory researchers are seeing a widespread and severe low-snow and low-runoff conditions across the region. Their modeling indicates the snowpack has peaked roughly a month earlier than usual and has begun to melt sooner than normal in the upper Colorado River Basin. A new study states that: "Climate change is altering the baseline conditions toward a continuing dryer state in the West, and that means the present conditions keeps getting graver," said the study's lead author Park Williams, a climate hydrologist at UCLA. "Climate change is literally baking the water supply and forest of the Southwest and it could get a lot more serious if we don't slow down the warming."

The Pacific Northwest

According to Oregon's Fifth Climate Assessment, the states average annual temperature has warmed by about 2.2 degrees Fahrenheit per century since 1895. More than a third of the state, on average, has been in drought since 2000. In Idaho, 58% of its area is experiencing moderate to exceptional drought conditions. The state's water resource department issued an emergency drought declaration in 34 out of its 44 counties in April. Glaciers in Washington's Olympic National Park could be gone by 2070 with permanent impacts on an important source of summer water for people and fisheries, according to a new study published in the Journal Geophysical Research, Earth Science.

California

Despite the dry conditions, urban water use in the state rose by nearly 19% in March of 2022. Over 6 million people in Southern California will have outdoor watering restrictions for the first time this summer, as the Metropolitan Water District orders outdoor watering only once a week in some densely populated cities. The cost of water for its citizens will continue to rise as the drought continues. In 2021 the ongoing drought cost thousands of jobs and over one billion dollars in lost production in the San Joaquin Valley's agriculture area; Hundreds of wells went dry and more are expected to dry up this year. California's two largest reservoirs, Lake Shasta and Lake Oroville, are at critical low water levels.

The Southwest U.S. is basically a desert. The states of Utah, Nevada, Arizona and much of Southern California could never had developed and prospered to what they are today without two things: Dams and a vast system of water distribution projects. Without these endeavors what would it be like and if they were lost how long would it take to revert back to their natural state? In all of earth's history only one desert civilization, out of dozens that grew up in antiquity has survived uninterrupted into modern times; Egypt, and its method of irrigation was fundamentally different than the rest. The civilizations of Assyria, Carthage, Mesopotamia, Incas, Aztec and Hohokam all disappeared, many by lack of water and in some cases due to salt build up in the water and soils they farmed.

Agriculture: Prices for vegetables have almost doubled since 2021 after the states that grow fresh produce saw water allotments reduced and fewer storms that decreased supplies. This November vegetable prices increased 38% from the previous month. On a year-to-year basis the price surge was more than 80%. As an example; a case of romaine lettuce that was priced last January at $25 is now at $75, a tripling of the wholesale price. Why? "There is just not enough water available to grow everything we normally grow," said Don Cameron, the president of the California State Board of Food and Agriculture. In California, the top agriculture producer in the U.S. experienced droughts this year that resulted in $3 billion in lost production due to the decrease in the amount of water available. These conditions have left tomatoes to wither on the vine and lettuce to shrivel in the fields. In Florida the hurricane Ian cost the produce industry 1.9 billion dollars in damage to the states agriculture industry, hitting oranges and tomato crops particularly hard.

All across the nation extreme weather has affected agriculture and many different crops. In California the counties of Colusa and Glenn, because of the lack available water, experienced a deep reduction of over 50% in their planted rice acreage. During 2022 in California seven percent of the state's croplands went unplanted due to lack of water. In Texas, a drought and heat wave reduced the harvest of cotton by nearly 70% in some areas. Extreme heat and lack of rainfall have severely damaged this year's cotton harvest in the U.S. which produces about 35 % of the worlds crop. The cotton growing region normally receives around 18 to 20 inches of rain per year. This year it saw less than 3 inches from August 2021 through the summer as nearly all of Texas baked under drought and blistering temperatures. Farmers chose not to harvest their fields because the plant production was so poor that it wasn't worth the cost.

In Georgia farmers have started growing citrus as the weather warms up and it becomes too hot to grow peaches. Also in Texas, the drought is forcing more ranchers to send cows to slaughter early. "There isn't enough grass available for cattle forage and it has become too expensive to buy feed. There has been a large amount of culling this year," said David Anderson a livestock specialist at Texas A&M University. Beef slaughter is up 13% nationwide and in the Texas cattle country it is up by 30%. The U.S. produces 82% of the world's almonds, nearly all in California. In 2022 the harvest was down 11% from the year before forcing some farmers to quit the business because of the lack of water for their trees; whole orchards are being removed. The year's production is expected to drop as much as 2.6 million pounds.

For the second year in a row, Arizona and Nevada will face cuts in the amount of water they can draw from the Colorado River as the West endures a continuing drought. In August of this year the Federal Government agency that regulates the water distribution from the Colorado River announced that Arizona's water deliveries will be reduced by 20% starting in January 2023. Arizona produces a large percentage of the country's winter supply of lettuce. The cuts to be made will force states to make decisions about where to reduce consumption and whether to select growing cities or agriculture for the reductions.

But those reductions represent just a fraction of the potential pain that may come for 40 million Americans in seven states that rely on the rivers water for drinking, electricity and agriculture. Because the states have failed to agree on the quantity of cuts to be shared, the federal water authorities may have to step in and make the 15% to 30% reductions that the government has said is needed the prevent the reservoirs from falling so low that they cannot be pumped. Lake Mead has reached a point where it is coming very close to a point where it would be a dead pool, unable to produce hydropower at Hoover Dam. The same is true for Lake Powell where if present conditions continue, it may not be possible to produce power from Glen Canyon Dam as early as 2023.

In June of 2022 the Arizona Republic newspaper published an investigation about how a Saudi Arabian agriculture company completed a deal with the state to permit it to access Phoenix's reserve water supply without limits, competition or fair compensation. The company Fondomonte, pumps in the face of a serious water crisis, an estimated 6 million gallons of water from the aquifer annually; this is enough to supply 54,000 single homes for a year. This water could be worth $4 million a year, but the company pays the state only $86,000 annually to lease the land. Hopefully there will be an effort by the state to force Fondomonte to pay the state as much as $38 million for the water it has used and profited from.

Chili announced an unprecedented plan to start water rationing in the capital of Santiago as the country suffers through its 12th consecutive year of drought. The governor of the city said "we're in an extraordinary situation that has never occurred in Santiago's 491-year history; we have to prepare that there might not be enough water for everyone who lives here." Rotating water cuts of up to 24 hours could soon be made for its 1.7 million residents. He added that the trend of less rainfall, mountain snow and early runoff is here to stay.

New Jersey, Trenton. The state is moving closer to drought status as reservoirs levels, stream flows and groundwater have all declined sharply this summer due to record heat and low rainfall.

From the dry and low water levels in Spain to the falling water levels of the major rivers like the Danube, the Rhine and the Po an unrelenting drought is afflicting nearly half of all Europe.

There has been no significant rainfall for almost two months in the continent's western, central and southern regions. Due to these low water levels the operation of one of Spain's largest hydropower plants has been shut down. The water level was below 23% of capacity before the shutdown. The plant first opened in 1966 and has never been shut down before. Is this the future of the rapidly receding Lake Powell and Lake Mead? According to a two- year projection of the Colorado River system from the Bureau of Reclamation and the U.S. water resource management office, Lake Mead's elevation could reach 992 feet elevation: it is currently at 1,045 feet which is only 24% of its normal capacity. This is the Bureau's "probable minimum" level the lake could reach within 24 months. If the lake reaches that dead pool elevation it could be a catastrophe.

Africa: The death of 7 million livestock due to the worst East African drought in four decades is causing a humanin disaster to occur. The Worlds Health Organization warns that more than 80 million people in Ethiopia, Kenya and Somalia now suffer from food insecurity. Beyond the loss of livestock vast areas of croplands have been parched and whole communities have been disrupted as families migrate in search of food, water and grazing lands.

Major rivers continue to dry up in Europe, Western U.S. and the Orient. China's longest and largest river the Yangtze, is experiencing the country's worst heat wave on record with some areas reaching 113 degrees Fahrenheit and the river is reaching its lowest level ever recorded. Crop production has been severely threatened by the low water levels and difficulties in river shipments of crops and commodities has created conditions of a state of emergencies in some areas in the most populated country in the world. The Yangtze River basin, that includes parts of 19 provinces, produces 45% of China's economic output according to the World Bank. Factories in China's southwest have shut down and cities have imposed rolling blackouts after reservoirs that generate hydropower ran low in a worsening drought. Companies in Sichuan province closed down or reduced production after they were ordered to ration power. Water levels at hydropower reservoirs are down by as much as 50%.

Recently the Bulletin of Atomic Scientist published a warning of what lies ahead for Europe in the near future. The problem is overwhelming according to Alexandre Tull and Elnath Eltahir, authors of the report. Much of Europe has been in the grip of a powerful drought not seen in over 500 years that has damaged natural ecosystems, crops and many sectors of the economy. Water availability has been severely limited with drinking water having to be trucked in to more than 100 towns in France. Many European rivers are hitting new record low water levels with major implications on crops, power production and shipping. Italy's Poe River, who's basin accounts for 40% of the country's agriculture production was so low it jeopardized the rice harvest. The Loire River in France could be crossed on foot in certain locations. High river water temperatures have forced power plants to lower production and have also reduced oxygen levels threating fish populations. Yields on major crops are projected to fall by at least 10-20 percent due to lack of rainfall and water restrictions. Saltwater intrusion has happened in aquifers in some areas near the coast as ground water has been depleted.

Industry experts and authorities predict that Spain's fall olive harvest will be nearly half the size of last year's production, another casualty of global weather shifts caused by climate change. High temperatures in May killed many of the blossoms on the trees in the Spanish orchards; olive trees cover 6.8 million acres across the Spanish countryside. The ones that survived produced fruits that were small and thin skinned because of the lack of rain. This year has been third driest in Spain since records were started in 1964. The country also had its hottest summer on record. Spain's 350,000 olive farmers typically harvest their crops in early fall but this year with the olives too small to pick many farmers left most of the fruit on the trees.

The predictions for production this year is 882,000 tons compared to 1.62 million tons in 2021. The smaller harvest has increased prices according to Filippo Berio a farm representative. The price for European olives to produce extra virgin oil has soured from $495 to $4,938 per ton. It is feared that the most widespread variety of olive trees cultivated throughout the region won't be able to adopt to such rapid change in climate.

Wildfires have become a major problem due to the drought and high temperatures. As of August 27, 2022 approximately 1,645,000 acres had been burned across the European Union. This is three times the average loss for this time of year. The worst burning has been in Spain, Portugal, Romania and France. Certain areas in Europe are on the path to becoming a desert.

Summer heat waves have become more frequent and intense. This frequency of heat waves is growing rapidly and they are lasting longer. Changes that have occurred in the earth's atmosphere due to the increase of carbon dioxide and methane have disrupted the balance of the Mediterranean Sea researchers have concluded. They note that there has already been significant lower precipitation and poorer crop yields in the lower areas that depend on glacier melt water for irrigation and human consumption.

By the early summer the Iberian Peninsula, Morocco and Algeria were suffering from a mega drought. The situation was extremely dire as fresh water was already in short supply following three years of low rainfall. This led to severely reduced snowpack, the source of water for summer flows of the Rhine and Po rivers.

The report ended with sobering predictions. "The European heat waves of 2022 were exceptional in magnitude and duration and would have been highly unlikely without human-caused climate change. With future warming inevitable as long as greenhouse gases continue to be released, in less than a couple of decades, the 2022 summer heat might well become the norm." the researchers concluded. "That Europe, one of the most developed regions on the planet, is already struggling to withstand today's extreme summer climate should raise a glaring red flag – that the time for lowering emissions is now or never."

The latest update from the U.S. Drought Monitor showed the major contrast between the extensive rainfall in the Northwest and the hot dry Southwest, both occurring during the same period. The report stated; "this feast or famine contrast is a pattern that the climate crisis tends to amplify creating severe swings from one extreme to another."

Social Disorder

"Only two things are infinite, the universe and human stupidity and I am not sure about the former." Albert Einstein

In parallel to the perils of climate change are the disruptions occurring today in our society that are threating the very foundations of our country and the freedoms we have always cherished, fought and died for. Never in the nations recent history have we been faced with the amount of fragmentation between our people that we are now experiencing. We have become a people so divided and so frustrated by extreme ideologies and yes, even hate, that we have ceased to exist as a nation united. Why? When and how did this start; this inability to think, to feel, to express your own and to consider the other guy's opinions without threats and intimidations?

A movement that spreads conspiracy theories on social media and by conservative media pundits fueled on Fox News and elsewhere is instigating actual threats and violence against law enforcement, the FBI, school boards, teachers, election officials and even libraires. They have shown up at schoolboard and election committee meetings wearing their body armor and totting their AK-15s and yelling threats. They continue to place repeated threating telephone calls to these dedicated volunteers who give their free time and energies working to ensure that our democracy functions. The animosity and hateful rhetoric on social media and conservative TV are directed particularly against the Federal Government in general and the FBI specifically. Threats and actual physical attacks against FBI facilities and personnel have occurred and the mobilization to potential violence by these extremist groups is increasing.

The head of the FBI, Christopher Wray, a Trump appointee recently told a Senate Judiciary Committee meeting that since the spring of 2020 investigations of violent domestic extremism have more than doubled. This clearly shows that when everyone from conservative members of Congress, Fox News characters and the extremist on social media spread conspiracy theories how dangerous it can become. They should realize the consequences of their actions quit their lying and shut up. Will they? If any FBI personnel are attacked, killed or injured the blame will be directly related to their foolish idiocy. The mission of the FBI to keep the public and the country safe is being jeopardized if that agency must spend time and effort protecting its own agents, employees and families against these right-wing extremists who themselves enjoy that same protection.

In 2020 there were 19,384 people killed by gun violence, a 35% increase over 2019.

Throughout the former president's time in office the far-right domestic extremist were encouraged by his criticism of anti-fascist activities and Black Lives Matter protesters as an endorsement of their actions. They were able to enjoy a measure of his protection as long as they were targeting the enemies that Trump designated. The January 6th insurrection on the Capital building and Trump's loss of the election have forced the extremist to change their tactics. They now vent their anger and attacks against all levels of our society.

Those who can make you believe absurdities can make you commit atrocities. Voltaire

It is a fact that a small percentage of our people; the ones with the big mouths, big guns, big egos and an overabundance of stupidity have reached a point that they believe that they should, without restraints, be able to dictate their agenda on the rest of society with threats, fears, coercions, extortions, and bullying, even at the point of a gun. When in history have, we experienced this type of destructive force take over a society and what were the results?

The former notorious leader of the past, Hitler, told the world; **"It is easier to fool the masses with a big lie then a small one."** He convinced his people that they were the "master race" and then proceeded to lead them into a conflict that set the world afire and killed millions of people. The "Big Lies" are still being told today, every day; did we learn anything? The only difference in these packs of human animals that are roaming our streets today is they are not wearing brown shirts with swastikas but camouflaged bullet proof vest and carrying an AK-47. Their aim is the same as Hitlers bullies, disruption and the destruction of the social order.

Who are these people and how did they evolve into this disruptive force in our society and where do they get their support and encouragement to flaunt the laws of this country?

There are people in high authority, Congressmen and even Congresswoman, Senators and an ex-President who swore an oath to defend and obey the principles this country was founded on who have ignored their sworn obligations, lied and abandoned the basic foundations that I am sure the majority of Americans believe in. They have flouted, laughed at and broken the laws of the country with no shame or care. They stand on their shaky platforms of irresponsibility and spill hate, disruption, lies and more lies but then preen on the public stage as if they were saints.

For what and why?

To pursue their own shallow self-interest for power and prestige they neglect the interest of their country. It is they that give aid and comfort to these disruptive forces and in some cases even openly encouraging their volatile behavior. A recent example: a

United States Senator, Lesley Graham said on Fox News, "If the former President (Trump) is indited (for his many crimes) there will be riots in the streets of America." That is how these destructive movements were born, matured and became a threat to the America we all love. They didn't just happen, they were encouraged and supported at the highest level of power and still are, but we can clean the slate and get rid of these pompous shallow individuals by the power we all still have: **VOTE!** If we leave the governing to others, others will govern possibly not in our favor or liking so we must always be vigilant.

"No man is above the law and no man is below it; nor do we ask any man's permission when we require him to obey it. Obedience to the law is demanded as a right; not asked as a favor."
Theodore Roosevelt

Law and order are a job for all of us. If we avoid it long enough, we will have anarchy and what we have built will be destroyed. It is like building a beautiful building and then turning lose a bunch of wild animals in it. This is the old war, the war of civilization against the barbarian, of peacefulness, order and hard work; against the heedless, the cruel and the destructive. Revolutionaries who practice violence simply want violence. They are unhappy with themselves and are so incapable of coping with reality that they try to disrupt it. They want violence to relieve their own anger and pent-up hatred.

What we have most to fear are not foreign interest but those within our own borders who think less of their country then of themselves, who are ambitious for power. Some of these individuals would subvert anything for their own political profit. They would twist the laws of their own country to obtain and maintain their control. Such men are always prepared to listen and follow a smooth-talking man to achieve and hold onto their own privilege and authority. This is what our country is experiencing today.

The following are excerpts taken from an article in the Helena Independent Record newspaper, written by David James.

In response to the recent FBI raid on Trumps residence, where hundreds of secrete and top-secret documents were discovered that he had stolen, the reaction from the right-wing extremist crowd and even members of congress have been an insult to the intelligence of the American people. This raid was carried out only after two attempts were made to convince Trump and his attorneys to return all of the stolen documents that were illegally taken from the White House. They were not his property; they were the property of the American people. When the last request was made to have them returned, one Trump's attorneys sign a document stating that all of the files had been returned. The FBI suspected and evidently had evidence that they were lying, that substantial numbers of top-secret files were still in his possession. Why would anyone suspect Trump of lying?

The FBI search warrant was then approved by a federal judge, the heads of the Justise Department and the FBI. What was found by the FBI should be disturbing to every American. Some of this country's most sensitive secrets were stored in an unsecured storage room; three secret files were found in Trump's desk, more were found in his closet and some were mixed up with newspapers, showing an obvious attempt to hide them. Among the thousands of items found were thirty-four (34) file folders marked top secret that were empty. Where are the contents of those folders and what did they contain. No one knows and Trump isn't telling. Who else has had access to these files in the eighteen months since they were pilfered? Could they have been photographed, stolen or sold to foreign interest?

The Washing Post just recently reported that some of the documents recovered from Trump's residence contained files of a foreign government's military defenses and its nuclear capabilities. The FBI recovered more than 11,000 government documents during its August 8 search according to the Post. Some of the documents detailed top secrete U.S. operations that require special clearances to view, not just top- clearance; and some were so restricted that even the current president's senior national security officials were not authorized to review them. The number of serious laws violated by Trump is beyond belief. He had no right to remove these files from government control or to endanger the security of the country and the many lives that may have been be put at risk if the information contained in the files were to fall into the wrong hands.

And what has been the Republicans and MAGA response to the action taken by the FBI to protect the interest of the country? On Truth Social (Trumps mouthpiece), Twitter and Facebook the hate speech about the FBI's effort has been off the wall. One extreme pundit wrote "This Means War." Stephen Bannon called the FBI "the Gestapo" and Michael Caputo stated "the FBI is the KGB." Dinesh D'Souza states the FBI is a gang "of dangerous criminals." Former Speaker of the House of Representatives, Newt Gingrich stated that "the FBI likely planted evidence against Trump." Sebastian Gorka, a former Trump adviser said the FBI was "declaring war." Mark Levin stated, "this is the worst attack on the republic in modern history, period. This is a Stalinist hunt."

"All people are shaken to their innermost core by an irresistible desire to submit to a strongman and at the same time hold sway over the defenseless. They are willing to kiss the shoes of a new master as long as they too are given someone to trample on." Benito Mussolini, Dictator of Italy, during World War Two.

And then there are those Republican lawmakers, like candidate for Governor of Arizona, Kari Lake, calling the FBI: "This illegitimate corrupt regime hates America and has weaponized the Federal Government to take down President Trump." Loen Boebert says this search is "Gestapo crap." Marjorie Taylor Green is calling to defund the FBI. Elize Stefanik, "Joe Biden's FBI is weaponized against their opponents." Kevin Mc Carthy, the present minority leader of the House of Representatives accused the

FBI of "intolerable weaponized politics." Senator Ted Cruz called the FBI's action "an attack on our constitutional republic." This is the Republican party of Trump today; how sad.

"A proper functioning Democracy depends on an informed electorate. If a nation expects to be ignorant and free, in a state of civilization it expects what never was and never will be. If we are to guard against ignorance and remain free it is the responsibility of every American to be informed." Thomas Jefferson

The former Department of Justice Inspector General, Michell Bromwich singled out both Graham and McCarthy by saying: "I have been dealing with law enforcement and criminal justice systems for 40 years. I have never experienced these types of hateful attacks against the FBI, DOJ lawyers and judges. He added "It's one thing for professional rabble rousers and liars such as Bannon to attack law enforcement the way they have, it's quite another for so called respectable political figures like Graham and McCarthy to do so. Their recent actions and words reflect that their politics are detached from reality, facts and principals."

"When politicians attack the conditional separation of power and seek to intimidate and marginalize a co-equal branch of judges and the courts; with absolute power, when the rule of law is trashed; then Democracy will spiral into the abyss of fascism. It worked for Hitler."—Retired Montana Supreme Court Justice James Nelson

Numerous death threats have been made by the "Mad Right" against all of those that may have been associated with the legitimate investigation and search. The Federal Judge who approved the warrant for the search of Trump's residence, after reviewing reams of evidence presented to him by the FBI and DOJ, received death threats after his name was cited in a press account. They have also attacked a man who has spent his entire professional life dedicated to the protection of the health and wellbeing of the American people, as if they themselves were experts in infectious disease; Dr. Fructi, as he plans to retire at the age of 81. President Biden, the FBI its agents, and employees, the judge and Democratic Congressional members, their wives and children have all been intimidated. The facts do not matter to these people, they have been so indoctrinated with the "Big Lie" and hateful smears for so long they have become oblivious to reality.

This toxic sickness in the social media networks and right-wing conservative news outlets has created an atmosphere of unrelenting attacks on the very foundations of our country. The present push towards totalitarianism in the country by the extreme right shows that the distinction between fact and fiction, between truth and falsehood no longer exist. It can only be changed by you, the voter. Do you vote for those individuals who have proven loyal to their country or those who continue to pursue this insanity? Do you listen to the "Big Lie" or do you ask question and get the facts? An indication of how the American public has reacted to the FBI is stated in a recent poll published in the USA Today paper (8/29/2022). Twenty two percent (22%) of Americans rate the FBI very unfavorably and fifteen percent (15%) somewhat unfavorably. The consequences of the misinformation and lies are effective; we have an uphill climb back to reality.

"The dummying down of America is most evident in the slow decay of substantive content in the enormously influential media, the 30 second sound bite (now down to 10 seconds or less) and the lowest common denominator programing. Credulous presentations on pseudoscience and superstition by, especially a kind of celebration of ignorance." Carl Sagan

In recent polls, American voters ranked "threats to democracy" as the most important issue facing the county. When during a time of climate change, inflation and a pandemic, this is a remarkable statement about the fragility of America's fundamental rights and freedoms. The country is seeing a large number of assaults on democracy from unpopular abortion bans to the banning of books, school restrictions against minorities and efforts to restrict the voting rights for millions of people. Politicians who spread lies and sought to overturn the results of the 2020 election are running for office that will put them in control of the country's election machinery. Meanwhile the Supreme Court is enforcing its own agenda on abortions, guns and the environmental protections that put them in opposition to the majority of the American people.

The following are excerpts from an article written by Norman Lear as he starts his second century; that appeared in USA Today. Norman is an Emmy- winning television producer and a co-founder of People for the American Way.

Menacing voices on the far- reaching tentacles of cable television and social media have captured and contaminated the minds of two many of our fellow Americans. Some of these individuals who were elected in the 2022 mid-term elections won because of the lies that were told, resentments that were nurtured, fears and big-otries that were inflamed by people who are poisoning our polit-ical climate to build their own power and influence. Information warfare is this century's battle front on authoritarianism. As we know all too well, minds that have been poisoned online can lead to bodies that carry out violence against others in real life. It is more than disturbing that these people who call themselves pa-triots embrace violence and drape themselves with American flags while they talk about a segregated nation even with talk about civil war. No matter how much they wave the flag their agenda is a dangerous one.

"When a man is drowned in a party (political party), plunged in it beyond his depth, he runneth a great hazard of being on ill terms with good sense of morality or both of them. Such a man can hardly be called be call a free agent, and for that reason he is very unfit to be trusted with people's liberty when he has given up his own." Britain's Lord Halifax, 300 years ago.

The Republican Party and its conservative ideas have been stolen by the dangerous bible thumping right-wing extremist. American conservatism used to mean and follow the ideas of the nation's Founding Fathers who rebelled against the tyranny of Great Britten but the party now embraces a tyranny of their own making that threatens the very foundations of our Democracy. Their beliefs about liberty, limited government and a respect for truth is no longer a priority. The "BIG LIE"; has become their major weapon to spread misinformation, rebellion and to justify their actions for reaching for more power and personal gratification. Such flagrant disregard for the Constitution before Trump and the election of 2016 would have been condemned by the Republican Party, but no more. They go out of their way to encourage it.

In any country today the empty factories, vacant warehouses and impoverished neighborhoods are the classic signposts of a civilization in slow terminal decline. In the history of the world, societies that became environmentally stressed, overpopulated or both developed the risk of getting politically stressed with eventually having their governments decline and eventually fail. When the people become desperate without any hope, they blame their governments that are not able to solve their problems. They try to emigrate, even at the peril of their lives, because their present locations have become unlivable. We see that happening today in several locations in the world. Eventually they fight and start civil wars. They think they have nothing to lose so they become terrorist, or support or tolerate terrorism. Are we tolerating terrorism today in America?

Civilizations die in familiar patterns. They exhaust natural resources and spawn parasitic elites who plunder and loot the institutions and systems that make a complete society possible and then the rot starts. The large urban centers die first, falling into irreversible decay. Authority unravels, artistic expression and intellectual inquiry are replaced by a new dark age, the triumph of crude spectacles and the celebration of crowd-pleasing stupidity.

Flooding, drought, souring temperatures, political dissention, pandemics, diseases, wars, the depletion of natural resources are cratering the foundations of the world's industrial society. When our industrial complexes increased in complexity, it made the world more vulnerable to collapse, not less. The soaring temperatures throughout all parts of the world are continuing to increase. Europe is baking in the worst drought in 500 years. Iraq and other countries in the Middle East, are suffering with 120 degrees heat that has damaged their infrastructure, caused serious water shortages and killed their people.

Agriculture production in most of Western, Central and Southern Europe is expected to decline by 8% by 2030. Power outages have become more frequent leading to the disruption of supply chains. This leads to the eventual slow destruction of the society we know. Civilization as we experience it today, is a very recent experiment in man's short history on the planet. We began our trip to the present several million years ago but it has only been about 10,000 years since we started down this modern path we are on. While the prevention of today's warming and the salvaging of the planet might have been possible fifty years ago, had we paid attention to the voices pleading for action; the possibility of a cure today might be impossible.

We humans don't seem to have the capacity to foresee, even given all the evidence, the magnitude of the changes that are occurring, or if we do we ignore them hoping they will go away. Unfortunately, this time they are not going away. This attitude might have been shaped by our history going back thousands of years when we lived day to day, struggling for existence when the only thing that was important was; "now."

Through lies and propaganda our social order today promotes greed and power in our large-scale societies that gives the rich and powerful a privileged vested interest in maintaining the status quo. By deceit they will continue to prosper long after the environment and the general public began to suffer.

"One of the saddest lessons in human history is this: If we have been bamboozled (lied too) enough we tend to reject any evidence of the bamboozle. We are no longer interested in finding the truth. The bamboozler has captured us. It is simply too painful to acknowledge even to ourselves, that we have been taken. Once

you give power to a charlatan over you, you almost never get it back." Carl Sagan

Our long history of attacking earths ecosystems to support our modern society has led to an "ecological deficit" where we have been developing, using and destroying more of Earth's biological assets then it can withstand. Nature will never be able to replace or repair, in the human time span, the damage we have inflicted. We have developed our energy, chemicals, and pesticides from earths 500-million-year-old carbon and distributed it throughout the world creating many ecological disasters for the coming generations of our species and all other living things. Some of those species have already perished; others will follow.

New Hampshire Bulletin, A review of a book by Eve Darian Smith.

There is a direct connection between the opposition to climate change policies and today's attack on our democratic institutions and the politicians and energy companies that support such efforts. It is a dangerous shift for both our representative government and the future climate. In recent decades our core democratic principles that prioritize citizens interest over corporate profits has been seriously undermined.

Today it is easy to find political leaders, on both sides of the political right and left, working on behalf of corporations in energy, finance, agribusiness, technology, military, gun and pharmaceutical companies that are not always in the public interest. These large multinational companies fund their political careers and election campaigns to keep them in office.

It's time to amend the Second Amendment to the Constitution: "The rights of the Voters Shall Not be Infringed." Is that any less important than the right to bear arms? Byrum

In the United States, this relationship was cemented by the Supreme Court's 2010 decision in the Citizens United case. This ruling essential said that corporations were people and allowed them and wealthy individuals to spend secretly, unlimited amounts of money to support the political candidates who best serve their interest. Donations to Republicans that came from the oil and gas industries more than doubled after this high court finding. More than one sixth of the money contributed to Democrats in the current political cycle as of March 31, 2022 went to Senator Joe Manchin, a strong supporter of the coal industry, and the only Democratic Senator that opposed the Presidents climate policy regarding coal emissions.

Since the 1990s, energy companies have heavily financed conservative candidates who support their interest and help to reduce regulations on the fossil fuel industry. This has enabled the expansion of fossil fuel production and use and increased CO_2 emissions to dangerous levels into the atmosphere. The industry's political power in shaping policy shows in the example of the 19

Republican State Attorney Generals and coal company's suing the Environmental Protection Agency from regulating greenhouse gas emissions from coal power plants, and by doing so, they prevented the regulations from taking effect. This was a real blow to our countries effort to reduce emissions and meet its commitment to lower the country's carbon footprint.

Oil companies' profits: During the third quarter of 2022 Exon-Mobil reported that it earned a net profit of $17.85 billion dollars and during the same quarter Chevron reported a net profit of $11.23 billion dollars, Shell reported $11.5 billion. This is during the time the American public was paying $4.80/ gallon of gas.

As an example of the influence of money can have on our political system spent by the energy company's; the following is the amount spent by the oil and gas companies during the 2020 election cycle to impact the election:

Koch Industries	$12,178,000
Shell Oil	$7,200,000
Occidental Petroleum	$7,010,000
Exon Mobil	$6,820,000
Chevron Corporation	$6,450,000

The amounts spent by the Electric Utility Companies:

Edison Electric	$9,974,000

Southern Co.	$9,100,000
American Electric Power	$6,969,000
Next Ere Energy	$6,205,000
Energy Capital Partners	$5,337,000

During the 2020 election cycle, the five biggest donner amounts from the oil and gas and coal industry went to conservative groups, Republican candidates, and the GOP. They were: Energy Transfer-$14,697,000; Koch Industries- $13,200,000; Chevron Corp. - $8,591,000; American Petroleum Institute- $5,201,000; Marathon Petroleum- $4,267,000.

At the same time that the energy industry has sought to influence policies on climate change, it has worked to undermine the public's understanding of climate science. Exxon Mobil participated in a widespread climate science denial campaign for years, spending more than $30 million on lobbyists, think tanks and researchers to promote climate-science skepticism. These same efforts continue today. A 2019 report found the five largest oil companies have spent over one billion dollars on misleading climate-related lobbying over the last three years.

In November of 1959, Edward Teller, the "scientist who developed the hydrogen bomb;" told a group of oil company executives and scientist gathered at Columbia University "that continued burning of fossil fuels would warm the planet, potentially melting the ice caps and submerging many coastal cities, posing a threat to civilization comparable to a global nuclear war." And now, after 64 years Teller's predictions are coming true. But instead of the big bang of instant annihilation we are being strangled slowly as the earth overheats. The final outcome may take longer but the end results may be the same.

During the late 1970s and into the 1980s, scientist at companies such as Exxon and Shell determined that the burning of fossil fuels was contributing to the rising of global temperatures. "There is concern among some scientific groups that once the effects are measurable, they may not be reversable and little could be done to correct the situation," a 1982 internal briefing for Exxon's management noted.

These corporate oil giants have known for a long time that their continued promotion of fossil fuels would threaten the very existence of mankind. Yet they have ignored repeated warnings and have continued to increase the production and promote their use while they practice "greenwashing" which is nothing more than lying by saying one thing to the American public and then doing the opposite. They preach the bogus idea of "carbon capture projects", such as planting trees and then turn around and try to secure tax credits for their fabricated efforts.

The energy industry has in effect controlled the political process and has prevented the enactment of effective climate policies.

Is it any wonder then that the country has been slow, sometimes to a crawl, to enact meaningful, sound and effective climate legislation like placing a tax on greenhouse gas emissions? It is critical that the people reclaim their democratic rights and vote out the anti-environmental political supporters, whose political power depends on the continued extraction of the earths limited natural resources over the best interest of their citizens and the future of humanity.

What does it mean when President Biden, in a speech at the 2022 World Cop 27 climate conference being held in Egypt, when he states that the U.S. is going to reduce its climate change emissions by 40 % by 2030; just eight years. Is it realistic or are we again treading water and going nowhere with our commitments? What would a 40 % reduction entail? Does it mean that 40% all of our cars, trucks, busses, boats, aircraft and houses will be burning 40 % less carbon than today? Does it mean we will be burning 40 % less oil, desal, natural gas, heating oil and coal in all the present day personal and industrial uses such as the production of plastic and energy? It is convenient to say we are going to achieve the 40% reduction by 2030, but it is extremely more difficult to say (and do) where these reductions are going to occur and whose "ox is going to be gored" in doing so.

Since 1959 the World has released 1.55 trillion tons of carbon into the atmosphere according to the Global Carbon Project scientist. In 2020, the last year when full national data was available, China released 11.7 billion tons of CO_2; that was 30.6% of the global emissions that year. This was twice the percentage the

U.S. released. The Europeans released 7.5 tons and India was at 7%. But scientists say that just looking at one year doesn't show who caused the long-range problem, because CO2 stays in the atmosphere for so many years. When looking back from 1959 to 2020, the U.S., not China is the biggest carbon polluter, it is not even close. The U.S had been the number one contributor over that time period, putting 334 billion tons of CO2 into the atmosphere, about 21.5 % of all the carbon released. This shows that poorer nations such as those in Africa had the least to do with causing climate change but are going to suffer the most in the future.

In the historical past, many individual societies collapsed but this time the collapse will be global. It will not be possible to move to new lands rich in natural resources. This will not be an option as much of the Earth's surface will become uninhabitable. And to believe that mankind will be saved by finding and inhabiting some distant planet is a pipe dream, it can never happen. The more insurmountable the crisis becomes, future societies will degenerate into self-denial, insurrection, wars and helplessness. Civilizations are not murdered they commit suicide. They fail to adapt and change as crisis insures their destruction. Our civilization's demise will be unique in size and magnitude caused eventually by our fossil fuel driven industrial culture.

"If destruction be our lot, we ourselves must be its author and finisher. As a nation of free men, we must live through all times or die by suicide." Abraham Lincoln

www.ingramcontent.com/pod-product-compliance
Lightning Source LLC
Chambersburg PA
CBHW060907140726
47996CB00001B/154